Manual of Carbonate Sedimentology: A Lexicographical Approach

Manual of Carbonate Sedimentology: A Lexicographical Approach

T. J. A. Reijers

Shell Internationale Petroleum
Maatschappij B. V.
The Hague, The Netherlands

K. J. Hsü

Geological Institute
Swiss Federal Institute of Technology
Zurich, Switzerland

With a foreword by K. J. Hsü

1986

ACADEMIC PRESS

Harcourt Brace Jovanovich, Publishers

London Orlando San Diego New York
Austin Montreal Sydney Tokyo Toronto

ACADEMIC PRESS INC. (LONDON) LTD.
24–28 Oval Road
LONDON NW1 7DX

United States Edition published by
ACADEMIC PRESS, INC.
Orlando, Florida 32887

British Library Cataloguing in Publication Data

Reijers, T. J. A.
Manual of carbonate sedimentology.
1. Rocks, Carbonate
I. Title II. Hsu, Kenneth J.
552'.5 QE471.15.C3

Library of Congress Cataloging in Publication Data

Reijers, T. J. A.
Manual of carbonate sedimentology.

Bibliography: p.
Includes index.
1. Rocks, Carbonate. 2. Sedimentation and deposition. I. Hsu, Kenneth J. (Kenneth Jinghwa), Date . II. Title.
QE471.15.C3R45 1985 552'.5 85-6098
ISBN 0–12–584840–4 (alk. paper)

PRINTED IN THE UNITED STATES OF AMERICA

86 87 88 89 9 8 7 6 5 4 3 2 1

Contents

3. Carbonate Diagenesis

4. Carbonate Reservoirs

Alphabetical Indices

Foreword

When I joined the Shell Exploration and Production Laboratory at Houston, Texas some 30 years ago the research laboratory was newly established by Royal Dutch and Shell Oil of America. Although applied research in such fields as chemical manufacturing and communications technology was then recognized as rewarding in terms of investment and earnings, applied geological research was largely unknown and its aims and objectives awaited clarification. The president of the company, a chemist, had experience in the Manhattan atom bomb project and expected a breakthrough in technology to yield a competitive advantage for oil search as the atomic bomb had in other respects at the end of World War II. Thus, to secure a large concentration of talent might well be a wise investment. So, it seemed to me, was the management rationale.

Research on carbonate rocks received a very high priority in the Houston research laboratory, and the Shell organization had recruited some of the most talented people available, including James Lee Wilson, Robert Ginsburg, Robert Dunham and many others who were to make a name for themselves in the profession. Aside from general studies of Recent and ancient carbonate sediments, the staff included specialists in paleontology (algae, rudistids, pelecypods, shell structures, etc.), sedimentology (marine environments, facies analysis, diagenesis, etc.), geochemistry (isotope geochemistry, chemical thermodynamics, radiometric dating, etc.), hydrodynamics (theoretical analysis, field measurements, analogue modelling), electron microscopy and petroleum engineering (porosity, permeability, capillary pressure tests, etc.). Shell also retained two of the most prominent sedimentologists, Francis Pettijohn and Robert Folk, as consultants. With some two dozen gifted scientists working for about a decade, the scientific achievements, as one might expect, have been extraordinary. However, simple-minded expectations of a "super-bomb" proved to be unrealistic; there were no easy or magic ways to find or produce oil. On the other hand, the research has demonstrated that knowledge pays and that scientific predictions could increase the success rate. After 10 years with Shell, many of us clearly recognized that oil exploration or production was not exactly our job description: our job was to do research in order to *help* geologists in the division or district offices to search for or to exploit underground petroleum reserves. After having obtained relevant results, it was our responsibility to communicate our knowledge to those who could make best use of it.

The communication between research scientists and operational staff took many forms. We visited regional offices, gave briefings to management and discussed technical details with operational staff. We gave training courses and seminars and prepared reports and manuals to be used by company geologists. It was almost exactly 20 years ago that I was given the task of synthesizing the results of carbonate research. It was not enough just to produce a summary of the good work being done by my brilliant colleagues since I was also asked to present, in easily understandable form, the relevance of the research results to exploration and production of petroleum reserves in carbonate rocks. A particular requirement was to produce a manual so that a geologist in the operational branch of the company might get started, might find a handle for predicting subsurface distributions of porosities and permeabilities in the carbonate terranes of the target area. The aim of a report I produced 20 years ago is the purpose of the present manual, now written largely by my coauthor, Dr. T. J. A. Reijers.

The first 10 years of pioneering research in Shell have carried us a long way. Some of us, at that time just out of school, had possibly heard of A. W. Grabau's distinguishing calcirudites from calcarenites from calcilutites, but to many others, a limestone was a limestone was a limestone. Studies of Recent and ancient carbonate in the 1950s have revealed the manifold origins of the particles and of the cementing materials in various carbonate rocks. We were taught systems of classifying carbonate rocks, one recommended by our consultant, Folk, another invented by our own staff, Dunham. We were shown the different sedimentary structures, bearing the characteristic signatures of various depositional environments, observable today in the Florida–Bahama region, or reconstructed from past carbonate terranes. Simulating the science of comparative anatomy of living organisms, a new science of comparative sedimentology was born, using Recent analogues to interpret the past. The next step was to relate the trend and geometry of carbonate sediment-bodies, which may be petroleum reservoirs, to their depositional environments.

Sediments are loose aggregates of particles, and diagenesis makes loose grains into rocks. Our ignorance of diagenesis in the early post-war years can be illustrated by an anecdote told by my friend Robert Ginsburg. As a graduate student he was sent from Chicago by Pettijohn to Florida to study the lithification of lime sediments. In a synopsis for his dissertation research, he proposed the use of boreholes to obtain samples from various depths where lime sediments were expected to show the various degrees of transition from loose agregates to solid rocks. His surprise was great when he found beach rock in the intertidal zone of south Florida, where lime sediments deposited "yesterday" are being lithified by cementation "today". He searched in vain for the postulated transitions. Piston cores from the deep sea may contain soft calcareous oozes many millions of years old, yet Pleistocene or Holocene sediments in the regions of active groundwater flow have largely become rock through vadose or phreatic diagenesis. Studies in diagenesis led to quantum jumps in our knowledge during the 10 years of "pioneering" research (1955–1965). Diagenetic textures and diagenetic structures were recognized, the chemistry and mineralogy of diagenetic minerals were clarified, and their stability in various diagenetic environments defined. The connection between hydrodynamic movements and diagenesis became clearer, and movement patterns of diagenetic solutions began to be used to predict the geometry of diagenetically altered bodies.

In an academic institution we would have been satisfied with the publications of our findings on the genesis and diagenesis of carbonate rocks, but working in industry, it was our task to apply these insights to practical problems. Oil underground is stored in the pores of carbonate rocks, and oil may be produced if the pores are sufficiently well connected (or if the host rock is sufficiently permeable) to enable its migration to the well head. Carbonate porosity and permeability may not be sufficient conditions for oil accumulation,—there should also have been source rocks and trapping—but a rock with little porosity and negligible permeability rarely makes a hydrocarbon reservoir. The purpose of the carbonate manual was, therefore, to enumerate the various ways in which a carbonate rock can be porous and permeable, and to discuss the various known types of petroleum-bearing carbonates. Through a correlation of the porosity types to the genesis and/or diagenesis of the porous rocks, some rules on the general patterns of porosity distribution could be worked out to help scientists in the regional offices predict carbonate porosity in the subsurface.

The first edition of such a carbonate manual was hastily written 20 years ago. I never possessed a personal copy because the manual was out of date shortly after it was written. If my colleagues and I at Shell made a good beginning, the achievement by the profession, by my colleagues in the industry and in the academic organizations surpassed our modest initial success during the next 15 years (1965–1980). Assigned to other tasks I no longer could keep up with the progress in updating the manual. Nevertheless, the need to translate basic science into practical use remained, and my colleague Dr. Reijers from SIPM, took it upon himself to prepare the present completely revised manual.

Our general philosophies and our understanding of applied carbonate research are similar, but this new manual has adopted a different approach and includes so much new material that it bears hardly any resemblance to the original compiled by me 20 years ago. Rather than a long and involved text, we have adopted an encyclopedic, lexicographical organization. And although an alphabetic index is available, the entries are not listed alphabetically; they are grouped together and ordered thematically. If the entries, terms for entities or concepts, can be considered the vocabulary of carbonate geology, the linking texts are the grammatical instructions on the interrelations of the words or phrases in such a language. The advantage of our system is that we can get into a project with a few key word entries: if we are looking for a reef reservoir, we may look up *reef* in Chapter 2 and the various porosity entries in Chapter 4; we could find relationship to other entries (entities or concepts) and bibliographical references for deeper understanding. In a way this book can be compared to a manual for the identification of plants or animals. If we have an object that can be tentatively identified, the manual may help in its identification and in understanding its scientific (and/or practical) significance. This approach, as far as I know, is new to the earth sciences, and we hope it will prove useful in practice. The translation of each entry into four languages should make the manual more universally applicable, and we may attempt Russian and Chinese translations of the terms for future editions. I have already profited a great deal from my colleague's linguistic ability: a preliminary copy of the manual has enabled me to minimize the use of English terms in lectures that have to be given in German. Perhaps our colleagues in German, French, Dutch and Spanish universities could similarly profit from this manual.

K. J. HSÜ

Preface

Carbonate sedimentology is a comparatively young branch on the tree of geological disciplines, but in its short life it has developed dramatically. In the 1950s the first serious attempts were made to introduce classification systems specifically designed for carbonate rocks. *AAPG Memoir I* (W. E. Ham, ed., 1962) established a number of useful classification systems with which carbonates could be described and better understood, and since then the discipline of carbonate sedimentology has evolved rapidly. Much attention has been and continues to be focussed on carbonate sedimentology by oil companies as a result of the further delineation of, and new discoveries in, major oil producing carbonate provinces. Considerable carbonate sedimentological work remains to be done in the sphere of exploration and production in the Middle East, the Caribbean area, Canada, the North Sea, and the United States, as well as many other areas.

The increased availability of consistent descriptions of carbonate sequences leads gradually to a better understanding of many depositional and diagenetic features. This fact is reflected in a number of authoritative works [e.g., textbooks by R. G. C. Bathurst (1971) and J. D. Milliman (1974)]. A comprehensive overview of the various aspects of carbonate facies—with emphasis on the oil–geological aspects—has been published by L. J. Wilson (1975) and a carbonate reservoir case history book, edited by P. O. Roehl and Ph. W. Choquette, has been published (1985). A number of special publications of the American Association of Petroleum Geologists and the Society of Economic Petrologists and Mineralogists has focussed on special aspects of carbonates. Some of the milestone publications are *Reefs in Time and Space* [L. F. Laporte, (ed.), SEPM. spec. publ. 18, 1974], *Deep Water Carbonate Environments—Concepts and Depositional Models,* [H. E. Cook and P. Enos, *Studies in Geology* **5,** AAPG, 1977], *Concepts and Models of Dolomitization,* D. H. Zenger, J. B. Dunham, and R. L. Ethington, SEPM spec. publ. 28, 1980], and *Dolomitization and Limestone Diagenesis,* L. C. Pray and R. C. Murray, (eds.), SEPM spec. publ. 13, 1965].

Other relevant books focussing on special environments or specific carbonate areas of the world are *Tidal Deposits, a Casebook of Recent Examples and Fossil Counterparts* [R. N. Ginsburg, (ed.), Springer, 1975], *Sedimentation on the Modern Tidal Flats of Northwest Andros Island* [L. A. Hardie, (ed.), The Johns Hopkins University Press, 1977], *The Persian Gulf,* [B. H. Purser, (ed.), Springer, 1973], *Pelagic Sediments: On Land and Under the Sea,* [K. J. Hsü and H. C. Jenkyn, (eds.), spec. publ. 1 Intern. Assoc. Sedimentologists, 1974], and *The Geomorphology of the Great Barrier Reef Quarternary Development of Coral Reef,* (D. Hopley, 1982).

This listing is far from complete and is only meant to show the rapid focussing on details that took place and is continuing. Almost all of the above publications have come from English-speaking countries. With the wider dissemination of concepts and ideas on carbonates, geologists from other countries have felt a need for appropriate translations of these concepts. For German-speaking geologists this need is met by the excellent textbook *Mikrofazielle Untersuchungsmethoden von Kalke,* (E. Flügel, 1978) which incidentally has recently (1982) been translated into English (*Microfacies Analysis of Limestones*). For French-speaking geologists Elf-Aquitaine (1975) issued a bilingual *Sèdimentaire des Carbonates/Sedimentology of Carbonates*. This work too, has recently been updated and reissued in an English version entitled *Exploration for Carbonate Petroleum Reservoirs,* (A. Reichmann and G. M. Friedman, 1982). The emphasis in both books is topical; however, the former concentrates on microfacies aspects whereas the latter emphasises a number of excellent hydrocarbon case studies.

What is lacking among the many publications on carbonates is a concise overview of the entire field of carbonate sedimentology with special emphasis on trends of porosity development; such a work would serve as a reference book for workers in the industry as well as for students in petroleum geology. The present book aims at filling that gap. It contains a comprehensive list of approximately five hundred key words along with their definition and translation into Dutch, German, Spanish and French. Moreover, widely used informal abbreviations are given. The increasing use of computers led to the addition of a number of alphanumeric codes for certain words or concepts. In the usage of both the informal abbreviations and the computer codes Shell's *Standard Legend* is followed. All defined terms, the abbreviations, and the computer codes are listed alphabetically at the end of the book in registers for the various languages. Quite a number of terms are illustrated by a standardised symbol, and it is suggested that one use such symbols for graphic representation in stratigraphic columns, cross-sections and/or facies maps. Such symbols are also useful in seismostratigraphic analyses of seismic sections. Shell's *Standard Legend* was here also used as a guide, but for a number of concepts no symbols exist and, where relevant, new symbols are proposed.

In searching literature and checking definitions the senior author came across an earlier, unpublished work by Dr. K. J. Hsü. It presented an overview of carbonate sedimentology that is still very relevant and only required updating. Relevant sections of this work were revised and as such integrated within some of the introductory sections. Dr. Hsü therefore accepted an invitation to co-author the present work and offered many constructive suggestions. Such suggestions, together with corrections of and additions to the various languages were also given by Dr. I. Zamarreño, Instituto "Jaime Almera" de Investigaciones Geologicas, Barcelona (Spanish), Dr. P. Freytet, Laboratoire de Géologie, Université de Paris Sud (French), Professor Dr. E. Flügel, Institut Für Paläontologie, Erlangen Universität (German) and Dr. J. D. Marshall, Geology Department, University of Liverpool (English). Dr. J. D. Marshall had the opportunity to comment on only part of the text.

A number of colleagues and friends both within Shell and from the academic world have also commented on portions of the work or provided some other useful information. My thanks go to all of them for their enthusiastic contribution.

Thanks are also expressed to the many publishing houses and authors who granted permission to reproduce illustrations from their works. They are acknowledged in the captions to illustrations throughout the book. Thanks are also due to Mr. E. F. v.d. Mortel who, with his typing staff, took care of punctual delivery of the many portions of typescript. Mr. A. P. v. Veen coordinated the reproduction of many illustrations and drew several new illustrations.

A special word of thanks is due to the management of Shell Internationale Petroleum Maatschappij who allowed the senior author to embark on this project and to use technical facilities of Shell as well as Shell internal information and who granted permission to release this book. Final responsibility for the entire work remains with the senior author, and the views expressed do not necessarily reflect the views of the Shell Group of Companies.

T. J. A. REIJERS

Explanatory Notes

This book is thematically subdivided. The various themes are collected within four chapters: (1)—Elements of Carbonates; (2)—Carbonate Environments, Settings and Bodies; (3)—Carbonate Diagenesis; and (4)—Carbonate Reservoirs. These chapters aim at (1) an acquisition of greater insight into the various elements that make a carbonate rock; (2) an understanding of where and how they are deposited; (3) a comprehension of how they are modified; and (4) an apprehension of how reservoirs are made out of carbonate rocks.

The chapters are subdivided into sections, as indicated in the list of contents. The subdivision was designed with the needs of the industry in mind. It enables one to find a required definition quickly within a logical thematic context. It also allows a quick overview, not only of the definition immediately required but also of related definitions. Where needed, cross references to related themes in other sections of the book are given.

In addition, a term can be located alphabetically in one of the language indexes at the end of the book. There, it is cross-referred to the page on which it appears and to its individual identifying number.

A number of sections in this book are introduced by short explanatory texts. Such texts are followed by the defined terms and their translations. The standard sequence in which the terms are defined is: Dutch (NE), German (DE), Spanish (ES), French (FR), computer code (RE) and abbreviation (AB). Not all translations can always be given.

During the search of the literature for definitions, it became apparent that certain terms are used in different ways. The definitions selected here are now the most universally used. Sometimes the preferred definitions remain arbitrary. It is realised that the meaning of certain terms develops in time with the acquisition of a better understanding of the process, product or concept defined. Definitions of terms aim at facilitating communication at this time.

The authors remain receptive to comments, criticism and additions by users of this book. Such comments can be sent to the publishers, who will send them to the authors.

List of Figures

1

Elements of Carbonates

1–1 CARBONATE SEDIMENTS

Introduction

Carbonate sediments belong to a group of deposits that are predominantly composed of chemical or biochemical precipitates. Such precipitates have a granular or, less commonly, a "primary" crystalline composition. The most important elements are

1. Whole or fragmented organic skeletons that become part of the sedimentary record as grains upon death of the organism
2. Grains and compound grains that are the result of direct precipitation or of organic binding
3. Crystals that form in the inter- and intragranular pore space and occlude this to various degrees
4. Microcrystalline carbonate mud (micrite)

Some carbonate sediments are made up predominantly or entirely of one type of grain. Frequently, however, carbonate sediments are a mixture of grains and some of the other elements just mentioned. Some precipitates (e.g., speleothems, travertine) produce directly a whole body of carbonate which is not granular in composition but entirely "primary crystalline". Mineralogically, carbonates are composed of such cations as calcium, magnesium and iron. The anion that is invariably present is CO_3^{2-}. The most common minerals encountered in carbonates are $CaCO_3$ (calcite, trigonal; or aragonite, orthorhombic) and $CaMg(CO_3)_2$ (dolomite, trigonal).

The accumulation of sediments, composed of grain- and mud-sized (division at 0.02 mm) particles, is governed mainly by physical processes in the depositional environment. Precipitation of crystals in the intergranular pore space (cementation) of the unconsolidated sediment produces a consolidated carbonate rock. (Bio)chemical processes may modify a carbonate rock to such a degree that it disintegrates and many of the constituents become unrecognisable. Chemical diagenetic processes may trigger replacement of deposited particles and the rock acquires a "secondary" (or diagenetic) crystalline aspect.

Dolomitisation is a very common diagenetic process. It is important from an oil geological point of view because 80% of the North American and ~50% of the world's carbonate reservoirs are dolomite. Until recently it was thought that directly precipitated dolomite did not occur; geochemical experiments and field observations have now cast some doubt on this belief (Baker and Kastner, 1981). The bulk of the Recent dolomites, however, are diagenetic in origin [see Zenger *et al.* (1980)]. Some are formed very soon after deposition, often while the sediment is in contact with surface-derived pore water. Advanced dolomitisation may transform the sediment into crystalline rocks bearing no trace of the original components.

Dolomite and calcite can be intermingled in various ways (Fig. 1–1):

1. Interbedded or strata-bound dolomite, which is often related to the depositional environment and the distribution of which is consequently somewhat predictable.
2. Irregular dolomite bodies that stand out in slight relief on weathered limestone surfaces, giving them a patchy appearance. This is often the result of selective dolomitisation of certain depositional textures or structures in carbonate rocks. The distribution of such a dolomite may also be predictable to a certain degree.
3. Uniform dolomite.

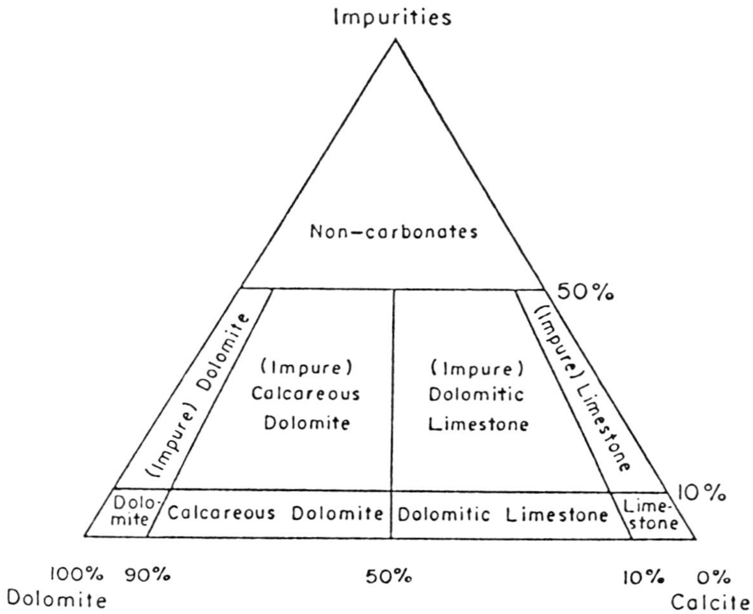

Fig. 1–1. Triangular diagram showing dolomite, calcite and carbonate admixtures.

Worldwide, limestones and dolomites are approximately equal in importance as reservoir rocks. Dolomites commonly have a higher permeability than limestones owing to their more uniform grain or crystal size.

Definitions 1–32

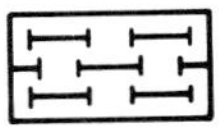

1. CARBONATE DEPOSIT; CARBONATE SEDIMENT: unconsolidated deposits formed *in situ* by organic or inorganic precipitation from aqueous solutions or by accumulation of detrital carbonate particles. The precipitates consist of carbonates of calcium, magnesium or iron.

NE afgezet karbonaatgesteente
DE Karbonatsediment; Karbonatablagerung
ES sedimento carbonatado; depósito carbonatado
FR dépôt carbonaté; sédiment carbonaté
AB carb. dep.; carb. sed.

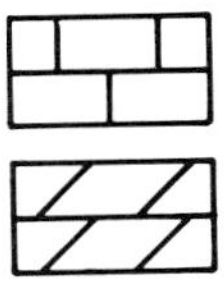

2. CARBONATE ROCK: rock composed of either cemented clastic carbonate particles or directly precipitated crystals with a bulk composition of more than 50% carbonate minerals (e.g., $CaCO_3$, $CaMg(CO_3)_2$).

NE karbonaatgesteente
DE Karbonatgestein
ES roca carbonatada
FR roche carbonatée

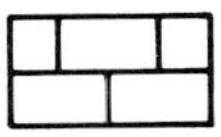 Ca

3. CALCITE: one of two crystallographic phases of calcium carbonate (trigonal). It is the stable phase in fresh water and in many diagenetic environments and will precipitate in preference to aragonite. Calcite is formed in many organisms whose skeletons contribute to limestone.

Note
1. Depending on the amount of magnesium carbonate present in calcite, one distinguishes high-magnesium calcite and low-magnesium calcite. The lower boundary is put at 8.0 mol % by Winland (1969).
2. The presence of magnesium carbonate makes calcite less stable. Other trace elements can also be substituted (e.g., Sr, Mn, Na, Fe).
3. For diagenetic calcite see Section 3–3.

NE calciet
DE Kalzit; calcit
ES calcita
FR calcite

RE CA
AB calc.

NE calciet, hoog magnesium
NE calciet, laag magnesium
DE Kalzit, hoch Magnesium
DE Kalzit, niedrig Magnesium
ES calcita, alta magnesium
ES calcita, baja magnesium
FR calcite faiblement magnésienne
FR calcite fortement magnésienne

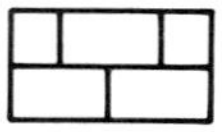
Ar

4. ARAGONITE: one of two crystallographic phases of calcium carbonate (orthorhombic). It is unstable when exposed to meteoric water. Nevertheless, it often precipitates from seawater in preference to calcite. This occurs, for example, when aragonite together with Mg-calcite is in contact with seawater. If the seawater is near saturation with respect to aragonite, it is supersaturated with respect to Mg-calcite and precipitates calcite while dissolution of aragonite continues. Both calcium carbonate crystallographic phases have a different solubility. Aragonite frequently contains appreciable amounts of trace elements, especially strontium.

Note: For diagenetic implications of the difference in solubility see Section 3–3.

NE aragoniet
DE Aragoniet
ES aragonit
FR aragonite

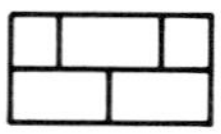

5. LIMESTONE: carbonate rock consisting almost entirely of calcium carbonate; specifically, a carbonate sedimentary rock containing more than 95% calcite and less than 5% dolomite.

NE kalksteen
DE Kalkstein
ES caliza
FR calcaire
RE LS
AB Lst (calc.)

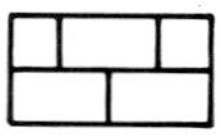
OC

6. ORTHOCHEMICAL LIMESTONE: a limestone originating from direct chemical precipitation.

Note: See Definition 341.

NE orthochemische kalksteen
DE orthochemischer Kalk(stein)
ES caliza ortoquímica
FR calcaire de précipitation chimique
RE LS

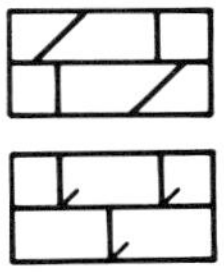

7. DOLOMITIC LIMESTONE: a limestone containing 10–50% dolomite and 50–90% calcite.

NE dolomitische kalksteen
DE dolomitischer Kalkstein
ES caliza dolomítica
FR calcaire dolomitique
RE DL
AB Dol Lst

Classification of Calcite-Dolomite Mixtures

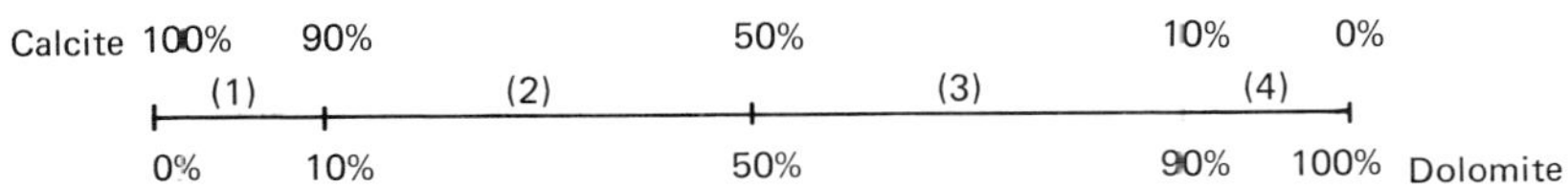

(1) Limestone, (2) Dolomitic limestone, (3) Calcitic dolomite, and (4) Dolomite

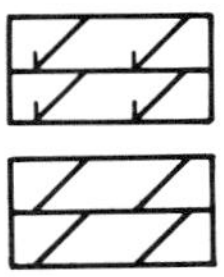

8. CALCITIC DOLOMITE (CALCAREOUS DOLOMITE): a dolomite rock containing 10–50% calcite and 50–90% dolomite.

Note: For calcite–dolomite mixtures see Definition 7; see also Fig. 1–1.

NE kalkige dolomiet
DE kalzitischer Dolomit(stein); kalkiger Dolomit(-stein)
ES dolomía calcárea
FR dolomie calcaire
RE DM LS
AB Dol. calc

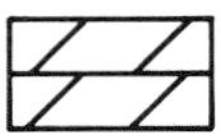 Dm

9. DOLOMITE (mineral): a common rock-forming trigonal mineral ($CaMg(CO_3)_2$).
DOLOMITE (sedimentary rock); DOLOSTONE (term not recommended): a carbonate sedimentary rock consisting predominantly (more than 50% by weight) of the mineral dolomite; specifically, a carbonate sedimentary rock containing more than 90% dolomite and less than 10% calcite.

Note
1. For calcite–dolomite mixtures see Definition 7 and Fig. 1–1.
2. For dolomitisation and dolomitisation processes see Definitions 326–332.
3. See also Definitions 374 and 375.
4. For textural terms see Definitions 81–86.

NE dolomiet
DE Dolomit (mineral); Dolomitstein (rock)
ES dolomita (mineral); dolomí (roca)
FR dolomite (minéral); dolomie (roche)
RE DM
AB Dol

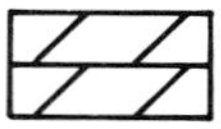
OC

10. ORTHODOLOMITE (PRIMARY DOLOMITE): dolomite formed by chemical precipitation in the depositional environment.

Note
1. The existence of fine primary dolomite under present-day circumstances is widely questioned, although Milliman (1974, p. 304) still suspects its existence, and more recently further indication of its existence has been found (M. Kastner, personal communication).
2. See also Definition 9 and Baker and Kastner (1981).

NE orthochemische (primaire) dolomiet
DE Orthodolomit; primärer Dolomit
ES ortodolomía; dolomía primaria
FR dolomie primaire
RE DM
AB Orthoch. Dol

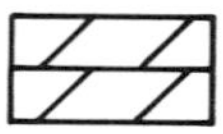
DP

11. PROTODOLOMITE: (1) a crystalline calcium magnesium carbonate with a disordered lattice in which the metallic ions occur in the same crystallographic layers instead of in alternate layers as in the dolomite mineral; (2) an imperfectly crystallised artificial material of composition near $CaMg(CO_3)_2$; (3) commonly thought to be a precursor precipitate to true dolomite.

Note
1. See also Gaines (1977), who defines protodolomite as follows: metastable single-phase rhombohedral carbonates which deviate from the composition of the dolomite that is stable in a given environment or are imperfectly ordered, or both, but which possess a high degree of cation order, as evidenced by the unambiguous presence of order reflections in diffraction patterns. See also "Discussion" on this matter by Deelman (1978).
2. See also Definitions 9 and 326.

NE protodolomiet
DE Protodolomit
ES protodolomita
FR protodolomite
RE DM
AB Proto Dol

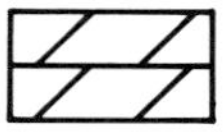
DC

12. CELLULAR DOLOMITE: dolomite with irregular, cellular or cavernous texture.

NE cellendolomiet
DE Zellendolomit
ES dolomía celular
FR dolomie cellulaire
RE DM
AB Dol

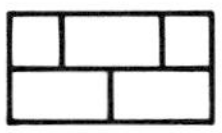
FC

13. FERROAN CALCITE: calcite with sufficient traces of ferrous iron in its crystal lattice to produce a stain in potassium ferricyanide which is different from that produced by pure calcite.

Note: See Evamy (1963) and Friedman (1959).

NE ijzerhoudende calciet
DE Eisenkalzit
ES ferroan-calcita
FR calcite ferrifère
RE CA
AB Calc.

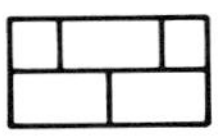
FD

14. FERROAN DOLOMITE: dolomite with sufficient traces of ferrous iron in its crystal lattices to produce a stain in potassium ferricyanide which is different from that produced by pure calcite.

Note: See Evamy (1963) and Friedman (1959).

NE ijzerhoudende dolomiet
DE Eisendolomit
ES ferroan-dolomita
FR dolomie ferrifère
RE DM
AB Dol.

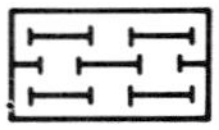

15. DREWITE (obsolete term): mud composed of needle-shaped aragonite crystals.

NE aragonietnaaldenslib; drewiet
DE Drewit; Aragonitnadel-Schlamm
ES drewita
FR boue aragonitique a cristaux aciculaires

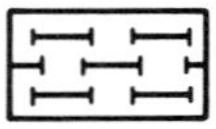

16. LIME MUD (CALCAREOUS MUD): unconsolidated clay-sized (up to 0.02 mm) carbonate sediment. Term does not have a precise connotation as to mineralogical composition.

Note
1. This definition follows Dunham (1962).
2. Compare with Definitions 89 and 127.

NE kalkslib
DE Kalkschlamm
ES fango carbonatado; barro carbonatado; barro calcáreo
FR boue calcaire; vase calcaire
RE LM

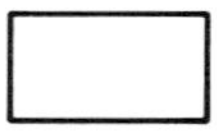

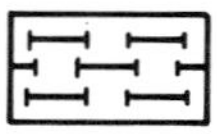

17. OOZE: (1) a soft, often semifluid mud (e.g., in rivers, estuaries and lakes); (2) a deep sea deposit, more than 30% of which consists of the remains of siliceous or calcareous pelagic organisms (see Fig. 1–2).

Note: See also Dean *et al.* (1985). They provide a comprehensive and updated classification system for fine-grained, deep-sea sediments.

NE slib
DE Schlick; Schlamm
ES lodo
FR vase; boue
RE LM

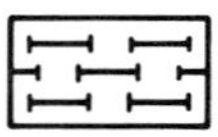

18. CALCAREOUS OOZE: a soft mud consisting predominantly of calcareous elements.

Note: See also Dean *et al.* (1985). They provide a comprehensive and updated classification system for fine-grained, deep-sea sediments.

NE kalkslib
DE Kalkschlamm
ES lodo cálcareo
FR vase calcaire; boue calcaire
RE LM; LS

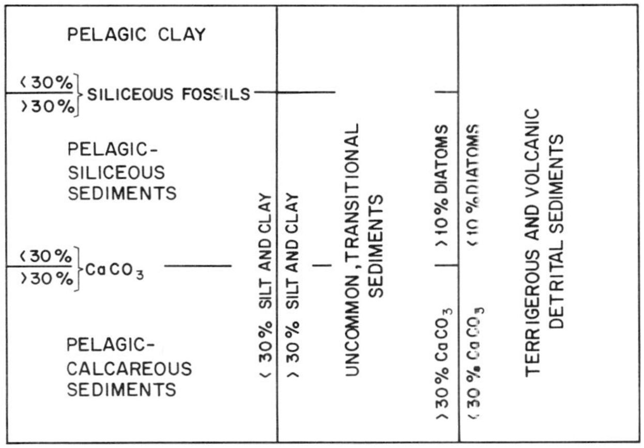

Fig. 1–2. Classification scheme for pelagic sediments.

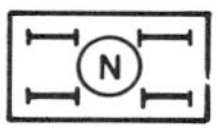

19. COCCOLITH OOZE: a soft mud, a very considerable part of which consists of skeletal remains of Coccolithophoroidea. Coccolith ooze is characteristic of deep marine environments.

Note: See Dean *et al.* (1985) and Definition 17.

NE coccolietenslib
DE Coccolithenschlamm; Coccolithen Schlamm
DE Coccolithenschlick; Coccolithen Schlick
ES lodo de cocolitofóridos
FR boue a coccolithes
RE LM NP

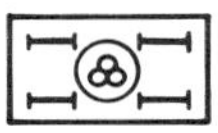

20. GLOBIGERINA OOZE: a soft mud, a very considerable part of which consists of skeletal remains (e.g., tests) of various planktonic Foraminifera. Globigerina ooze is characteristic of deep marine environments.

Note: See Dean *et al.* (1985) and Definition 17.

NE globigerinenslib
DE Globigerinenschlamm; Globigerinien Schlamm; Globigerinenschlick; Globigerinien Schlick
ES barro de globigerinas; lodo de globigerinas
FR boue a globigérines
RE LM FP

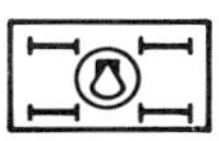

21. PTEROPOD OOZE: a soft mud, a very considerable part of which consists of shells of pteropods. Pteropod ooze is characteristic of deep marine environments.

Note: See Dean *et al.* (1985) and Definition 17.

NE pteropodenslib
DE Pteropodenschlamm; Pteropoden Schlamm; Pteropodenschlick; Pteropoden Schlick
ES barro de pterópodos; lodo de pterópodos
FR boue a ptéropodes
RE LM LP

22. CHALK, general (PRIMARY CHALK): fine-grained, finely porous, friable limestone composed to a large extent of coccolith fragments.

Note
1. The term is often mistakenly used for any white, friable limestone.
2. The "Chalk" is a specific lithostratigraphic term for Upper Cretaceous chalks of Northwest Europe.

NE oorspronkelijk krijt; kalk
DE Kreide; Schreibkreide
ES creta
FR craie
RE CK
AB Chk

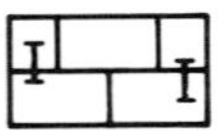

23. SECONDARY CHALK (CHALKIFIED LIMESTONE): Fine-grained, friable limestone produced by vadose leaching or dissolution of any pre-existing limestone.

Note: See also Definitions 25 and 308.

NE gevormd krijt
DE sekundärer Kreide
ES creta secundaria
FR craie secondaire; calcaire pulvérulent par altération
RE LSCK

24. CHALK ROCK (hard): indurated chalk formed as a result of ongoing cementation processes.

Note: See Mimram (1977).
Comment: Discussion of four stages in diagenetic development chalks and geochemical criteria.

NE krijtgesteente
DE Kreidegesteine
ES creta

FR craie durcie
RE CK
AB Chk

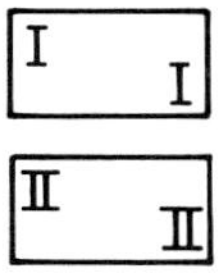

25. CHALKY: (1) having the appearance of chalk; (2) said of the porosity of such finely textured rocks as chalk and marl; (3) adjective describing admixtures or streaks and lenses of material described in Definition 22.

Note: See also Definitions 23, 25 and 430.

NE krijtachtig
DE kreidig
ES "chalky"
FR crayeux
RE CK

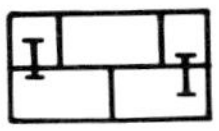

26. CRUMBLY LIMESTONE (FRIABLE LIMESTONE; CHALKY LIMESTONE): limestone from which the grains can be detached with a fingernail; rock crumbles between the fingers.

NE kruimelige kalksteen
DE zerbröckelnder Kalkstein; krümeliger Kalkstein
ES caliza friable
FR calcaire friable
RE LS CK
AB Lst chky

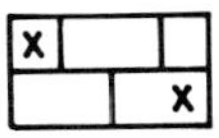

27. COMPACT LIMESTONE: hard and usually dense limestones

NE harde, dichte kalksteen
DE dichter Kalkstein, harter-, fester-, kompakter-
ES caliza compacta
FR calcaire compact
RE LS C7
AB Lst, cmt

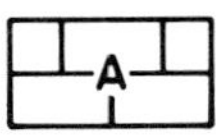

28. LITHOGRAPHIC LIMESTONE: aphanitic compact carbonate rock composed of uniformly sized, very fine grains or crystals.

Note: The term reflects the former use of such limestones in the graphic industry.

NE lithografische kalksteen
DE lithographischer Kalkstein

ES caliza litográfica
FR calcaire lithographique
RE LS A
AB Lst, aph.

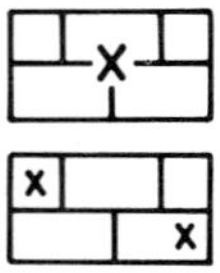

29. CRYSTALLINE LIMESTONE: limestone consisting predominantly or entirely of crystals.

NE kristallijne kalksteen
DE kristalliner Kalkstein
ES caliza cristalina
FR calcaire cristallin
RE LS X
AB Lst, xln

30. TUFA (CALCAREOUS TUFA; CALC TUFA): a spongy, porous carbonate rock that forms a thin surficial deposit near springs and rivers.

Note
1. Tufa forms by direct inorganic precipitation of calcium carbonate as a result of CO_2 release from water supersaturated with respect to calcium carbonate.
2. Tufa commonly includes remains of plants, etc. (including algae). The morphology of such inclusions could be adopted.
3. In French the term *tuf* can be used in a very wide sense, including all rocks that can be worked with a pickaxe. Consequently one could refer to tuf volcanique; craie tuffeau, etc.

NE kalkkorst
DE Kalktuff
ES toba
FR tuf calcaire

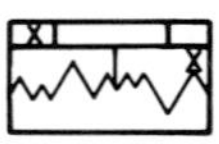

31. TRAVERTINE; CALCSINTER (the compact variety of tufa): a massive, often finely layered colloform type of chemical limestone, with or without porous interlayers.

NE travertijn
DE Kalktuff; Kalksinter; Travertin
ES travertino
FR travertin

32. SPELEOTHEM; DRIPSTONE: any secondary mineral deposit, but commonly calcite, formed in a cave or along a surface by precipitation from an aqueous solution. It includes stalactite, stalagmite and other laminated crusts.

Note: See Definition 345.

NE speleoliet; druipsteen; grotafzetting
DE Speleolit; Höhlenkarbonate; Sintergebilde; Speleothem
ES espeleotema; concreción
FR concrétion de cavité karstique; spéléolite

1–2 HOMOGENEOUS MIXTURES

Introduction

Limestones and dolomites, as discussed in the previous section, may grade imperceptibly into shaly, silty or sandy limestones or dolomites and from there into calcareous shales, siltstones or sandstones. This reflects an increasing proportion of insoluble siliciclastic material that can be present as either homogeneous or inhomogeneous admixtures. In the former case no further indication can be given of the way in which the clastic material is present, but when inhomogeneous admixtures occur, it is often possible to indicate the presence of streaks, interbeds or lenticular masses of such insoluble material. The division between "pure" rock types and homogeneous mixtures is usually placed at 10% insoluble material.

Definitions 33–39

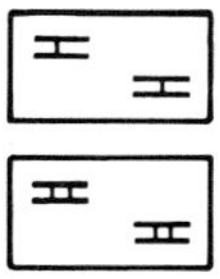

33. CALCAREOUS DEPOSIT: a deposit that contains up to 50% calcium carbonate.

NE kalkafzetting
DE Kalkablagerung
ES depósito calcáreo
FR dépôt calcaire
AB Calc. dep.

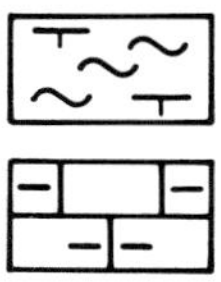

34. MARLSTONE; ARGILLACEOUS LIMESTONE: respectively friable or compact mixtures of clay material and carbonates (35–70% fine-grained carbonates and a complementary content of clay).

NE mergelsteen
DE Mergelstein
ES marga; caliza arcillosa

FR marne; calcaire argileux
RE AL
AB Mrlst; Arg Lst

Classification of Clay(stone)-Lime(stone) Mixtures

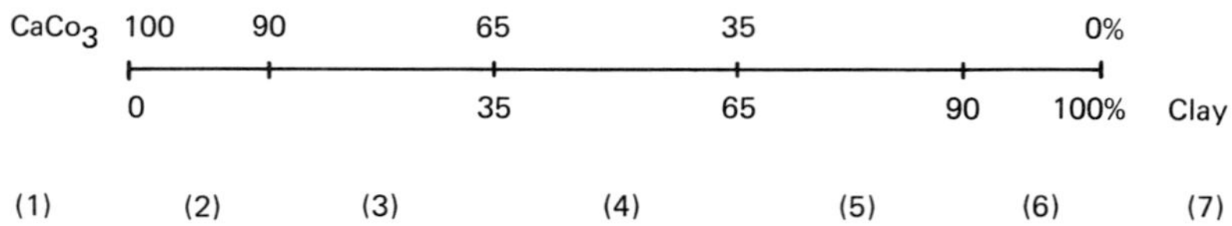

(1) Lime(stone), (2) Marly lime(stone), (3) Limey marl(stone), (4) Marl(stone), (5) Clayey marl(stone), (6) Marly clay(stone), (7) Clay(stone)

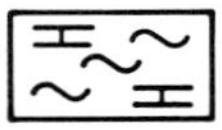

35. LIMEY MARL: a marl containing more than 65% carbonate but still an appreciable amount of clay (see Fig. 1–3).

Note: See Definition 34.

NE kalkige mergel
DE (stark)kalkhaltiger Mergel
ES marga calcárea
FR marne calcaire
RE MR LS
AB Mrl, calc

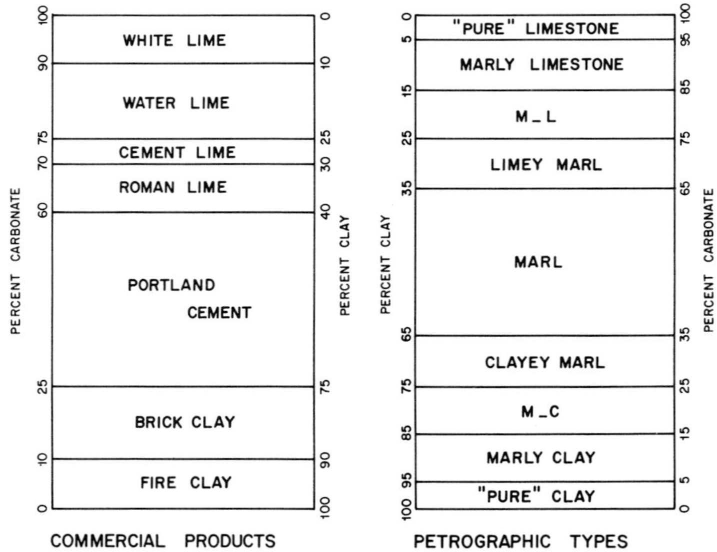

Fig. 1–3. Percentages of clay–lime carbonate mixtures as suggested by Barth *et al.* (1939) for commercial products and for petrographic types.

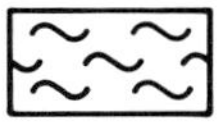

36. MARL: the friable equivalent of marlstone; a mixture of clay material and carbonates.

Note: See Definition 34.

NE mergel
DE Mergel
ES marga
FR marne
RE MR
AB Mrl

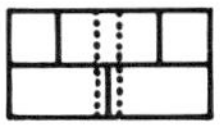

37. SANDY LIMESTONE: limestone homogeneously mixed with 10 to 50% sand of siliciclastic composition (See Fig. 1–9).

NE zandige kalk
DE sandiger Kalkstein
ES caliza arenosa
FR calcaire grèseux
RE LS SA
AB Lst, s.

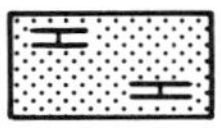

38. CALCAREOUS SAND: unconsolidated sand-sized clastic particles with a considerable admixture of carbonate particles (usually more than 20%) (See Fig. 1–9).

NE kalkzand
DE Kalksand; kalkreicher Sand
ES arena calcárea
FR sable calcaire
RE SA LS
AB S, calc

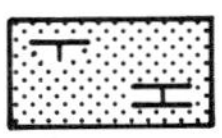

39. CALCAREOUS SANDSTONE: sandstone of a siliciclastic composition with a considerable amount (usually over 20%) of carbonate (See Fig. 1–9).

NE kalkzandsteen
DE Kalksandstein; Kalkreicher Sandstein
ES arenisca calcárea
FR grès calcaire
RE SS LS
AB Sst, calc

1–3 CARBONATE PARTICLES

Introduction

For practical reasons (visibility with hand lens with magnification 10×), Dunham (1962) divided carbonate sedimentary particles into two major groups on the basis of their size: particles less than and greater than 0.02 mm. In terms of origin they can be classified as particles with a skeletal, a chemical, a detrital or a mixed origin (e.g., compound particles). The origin of the fraction that is less than 0.02 mm (lime mud) cannot always be established with certainty. It appears that a considerable amount of this fraction has its origin in mechanical and/or organic breakdown. In the Recent, aragonite needles are produced in abundance by common green algae, and they provide a major amount of the lime mud fraction. In addition, some of these needles may be of direct chemical origin (precipitated from seawater), although the arguments for such a process (e.g., Cloud, 1962) have been contended. The origin of the fraction that is greater than 0.02 mm is mostly skeletal, detrital or compound (in that order of importance), and hand lens or binocular inspection often allows identification.

Skeletal Grains

Upon death, wide varieties of organisms (e.g., algae, corals, forams) contribute their skeletons to the sedimentary record. Such skeletons differ in size, shape, specific gravity, micro- and macrostructure and mineralogy. Due to these differences a range of sedimentary particles may be derived from various skeletons. In addition, selective depositional processes (organic and mechanical breakdown) eventually yield sediments of widely varying composition and texture.

A detailed understanding of the skeletal architecture of the various organisms is needed to determine with some confidence carbonate particles in ancient limestones (Majevske, 1969; Horowitz and Potter, 1971). The morphology of the skeleton determines its relative resistance to breakdown. The following series of skeletal morphologies offers increasing resistance to breakdown: (1) sheated or spiculed; (2) segmented, (3) branched; (4) chambered, (5) incrusted and (6) massive. Grouping organisms on the basis of skeletal morphology aids in evaluating the nature of the environment of deposition, in particular with respect to physical processes. But grain size alone is not a dependable index of variations in wave and current energy. Nor do sediments, derived from skeletons in recognisable preserved organic communities, necessarily reflect these communities perfectly. This is a direct consequence of the tendency in carbonates towards selective preservation of the more resistant grains and disintegration and removal of the fragile ones. Mineralogical variations in skeletal grains render them susceptible to different diagenetic processes (see Chapter 3, this book).

The distribution of marine organisms, and thus of particles and skeletons derived from such organisms, varies systematically with changes in environmental conditions. Amongst the specific factors controlling the distribution of skeletal particles it is often apparent that the variety and abundance of sessile organisms depend on unrestricted water circulation. There will be no attempt to define all the skeletal particles (bioclasts) in this work. Only some general terms will be defined; for detailed discussions of matters touched on above, reference is made to the textbooks mentioned in the Preface.

Detrital Grains

Grains resulting from the fragmentation of lithified sediments are referred to as detrital. One could question their sedimentary origin. It could be argued that they are diagenetic since they originate from sedimentary rock composed of particles that have reached the stage of final deposition (see Chapter 3).

Detrital grains can be of local origin, or they could have been brought to the depositional site from far away. Locally derived skeletal grains may indicate occasional high-energy conditions in the depositional environment. Detrital grains have been observed in tidal flat settings and in shallow marine sediments where lime muds, hardened by (repeated) exposure, are broken by erosion and drying, and swept into nearby settings. They then contribute to the accumulating sediment. It is often practical for descriptive reasons and necessary for environmental reconstructions to note the size, external shape and internal composition of detrital grains.

Compound Grains

Grains with a compound architecture or an uncertain, multigenetic origin are grouped in this category. They can be divided into ooids, pelletoids and pisoids, with a range of graduations between these end members. Many have a (semi)rounded shape in common. Intraclasts, although grouped as detrital grains, are commonly compound in internal architecture.

Ooids occur mainly where bottom sediments are in frequent motion in tropical climates. The restricted occurrence probably reflects hydraulic parameters controlling their precipitational mechanism. The difference in nature and abundance of Recent ooids can often be attributed to variations in conditions in the depositional environments, in particular the hydraulic energy. There seems to be a direct relation between high-energy conditions, thickness and crystal orientation of cortex and magnitude of ooid sand bodies (cf. Reijers and Ten Have, 1983).

Pelletoids include all discrete, structureless sand-sized carbonate grains that have regular outlines. A descriptive term is preferable to a genetic one, owing to the polygenetic origin of pelletoids (see Figs. 1–5 and 1–6).

Pisoids and oncoids are gradational to one another in size and perhaps occasionally in origin (algal). Pisoids, however, may also reflect a vadose diagenetic environment (Dunham, 1969a).

The arrangement of terms discussed hereafter will follow the outline just given. First some general terms will be defined. No individual skeletal grains will be discussed, but some attention will be directed to detrital grains and compound grains.

Definitions 40–71

40. PARTICLE: a general term indicating any carbonate fragment or grain. No specific size is implied.

NE deeltje
DE Partikel; Teilchen; Komponenten
ES partícula
FR particule

41. GRAIN: a specific term (Dunham, 1962) indicating a particle of more than 0.02 mm.

Note: See also Definition 42.

NE korrel
DE Korn
ES grano
FR grain

42. GRANULAR: (1) applied to crystallised or recrystallised sedimentary rocks when grains exceed 0.06 mm in size and are roughly equidimensional; (2) having a texture appearing to consist of grains.

Note
1. *Grains* is here used in a wider sense than as defined by Dunham (1962).
2. See also Definition 41.

NE korrelig
DE körnig
ES granular
FR granulaire

43. MICROGRANULAR: pertaining to carbonate grains or crystals between 0.001 and 0.06 mm in diameter and roughly equidimensional.

NE mikrokorrelig
DE mikrokörnig; feinkörnig
ES microgranular
FR microgranulaire

C LS DO

44. CLAST: a general term indicating an individual mineral, rock constituent, grain or fragment of a sediment or rock, enclosed in sedimentary material. More specifically it is frequently used to indicate a broken particle or shell fragment.

NE deeltje; fragment; brokstuk; klast
DE Bruchstück; Fragment; Klast
ES clasto; fragmento
FR débris; fragment

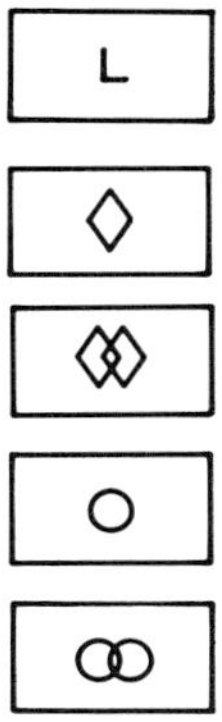

LS DO

45. LITHOCLAST: a fragment of rock embedded in subsequently formed sedimentary rock.

Note: The term *lithoclast* could be used for rock fragments of any composition, but it is commonly used for carbonate rock fragments.

NE gesteente fragment
DE Lithoklast; Gesteinsbruchstück
ES litoclasto
FR débris de roche; lithoclaste
RE LC
AB Lcl

NON-SKELETAL PARTICLES are primarily classified according to degree of rounding and aggregation:

◊	Angular fragment, lithoclast	○	Rounded particle (not determined further)
	Lithoclasts, aggregated		Rounded aggregated particles (grapestone)

L LS

46. LIMECLAST: a lithoclast made up of limestone that behaves as a detrital particle and has been produced by disintegration of a larger limestone mass either within or outside the depocenter.

NE kalksteenfragment
DE Kalkbruckstück; Kalkfragment
ES clasto calcáreo; fragmento calcáreo
FR élément calcaire (dans une brêche)
FR débris calcaire; grain calcaire (dans un grés)
RE LS LC

L DO

47. DOLOCLAST: a lithoclast made up of dolomite that behaves as a detrital particle and has been produced by disintegration of a larger dolomite mass either within or outside the depocenter.

NE dolomietfragment
DE Dolomitbruchstück; Dolomitfragment
ES clasto dolomítico; fragmento dolomítico
FR débris dolomitique; élement dolomitique
RE DM LC

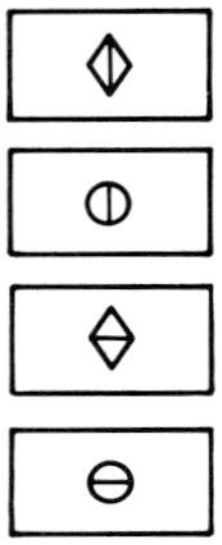

48. MICROPHENOCLAST: a clastic element smaller than 4 mm.

Note: The term is only occasionally used in various languages, and the concept can be indicated by the expression "grain of sand size".

NE mikrophenoklast
DE Mikrophenoklast
ES microfenoclasto
FR microphenoclaste

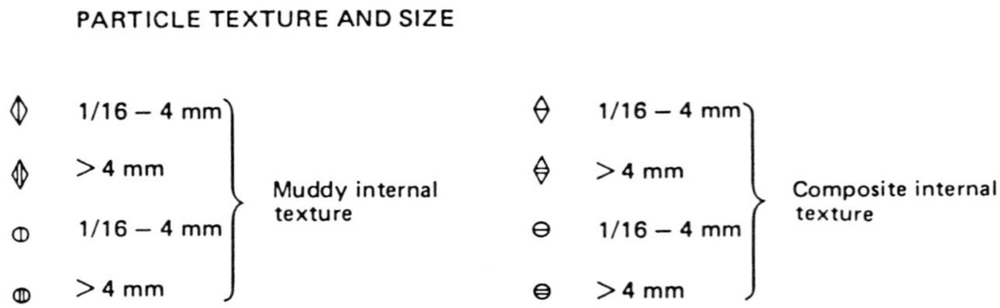

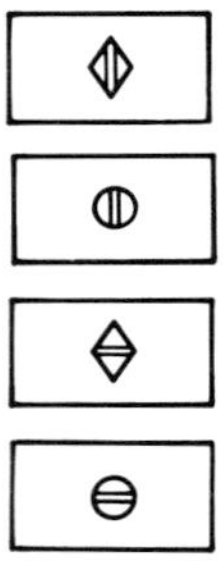

49. MACROPHENOCLAST: a clastic element larger than 4 mm.

Note: The term is only occasionally used in various languages, and the concept can be indicated by the expression "grain of granule or pebble size".

NE makrophenoklast
DE Makrophenoklast
ES fenoclasto
FR macrophenoclaste

I LS DO

50. INTRACLAST: fragment of penecontemporaneously, generally weakly consolidated carbonate sediment that has been eroded from adjoining parts of the sea bottom and redeposited within the area of original deposition.

NE authochtoon omwerkingsdeeltje
DE Intraklast; synsedimentäres Resediment
ES intraclasto
FR intraclaste; exoclaste

E LS DO

51. EXTRACLAST: a fragment of sedimentary material, produced by erosion of an older rock outside the area in which it accumulated.

NE allochthoon omwerkingsdeeltje
DE Extraklast; postsedimentäres Resediment

ES extraclasto
FR extraclaste

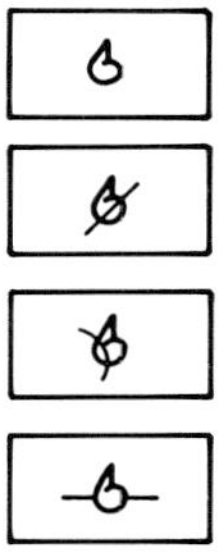

52. BIOCLAST: skeletal fragment (see Fig. 1–4).

NE bioklast
DE Bioklast
ES bioclasto
FR bioclaste
RE SF
AB Bcl

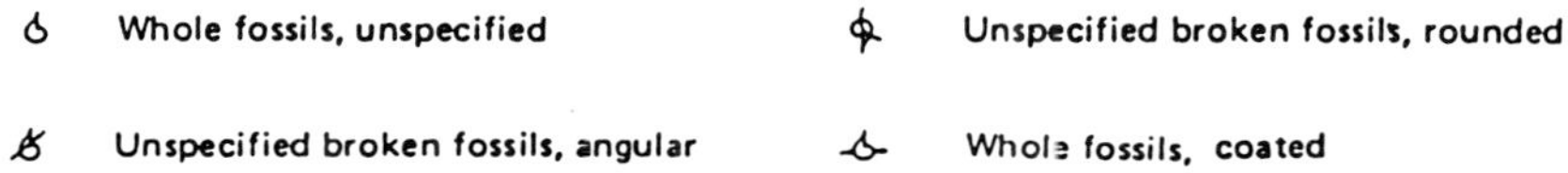

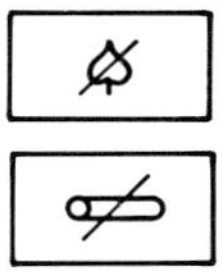

53. PHYTOCLAST: a transported remnant of a plant or tree that forms part of a sediment.

NE fytoclast
DE Phytoklast
ES fitoclasto
FR phytoclaste; débris végétal; phytólite
RE CL PL

54. SKELETAL: pertaining to hard remains of organisms; also said of rocks or rock materials that contain such remains.

NE skelet-, uit skeletresten bestaande
DE Organismenhartteile
ES esquelético
FR squelettique
RE SF
AB Skel.

55. ALLOCHEM: a collective term for one of several varieties of discrete and organised carbonate aggregates that serve as the coarse framework grains in most mechanically deposited carbonates; they include intra- and extraclasts, ooids, pellets, lumps and bioclasts.

NE allochemisch deeltje
DE Allochem; Karbonatkomponent

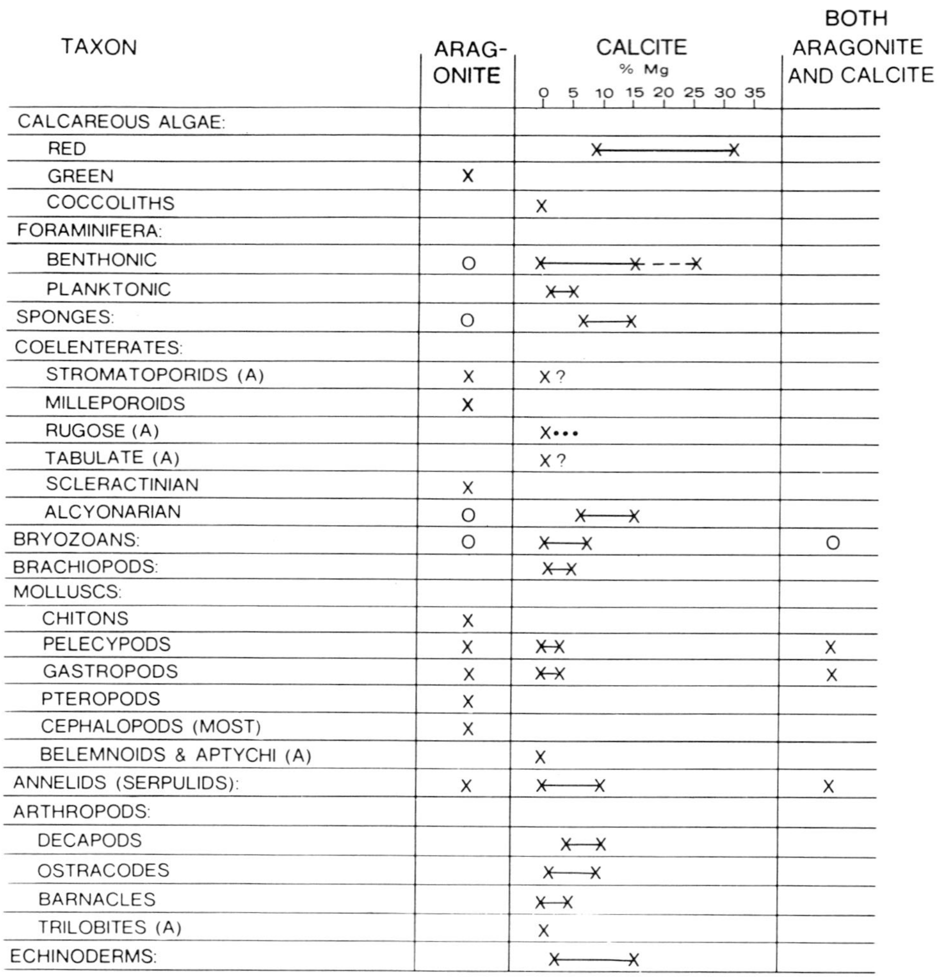

Fig. 1–4. Skeletal composition of major taxa in terms of carbonate mineralogy after Scholle (1978) by permission of the Am. Assoc. Pet. Geol. ×, Common; ○, rare. The letter A indicates that the composition is not based on modern forms.

ES aloquímico
FR allochème

56. ORTHOCHEM: a particle of carbonate that is formed by direct chemical precipitation within the depositional or the diagenetic environment.

NE orthochemisch deeltje
DE Orthochem

ES ortoquímico
FR orthochème

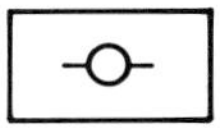

57. COATED GRAIN: a particle composed of a nucleus around which either (calcified) organic tissue or carbonate precipitate is present in one or more layers. Less commonly iron minerals also form a coating around a nucleus.

NE korrel omgeven door (verkalkte) organische, of precipitatie laagje(s); omhulde korrel
DE Rindenkörner
ES grano revestido
FR grain revêtu
RE OD

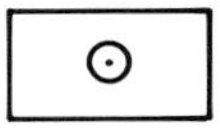

58. OOID: a spherical or subspherical accretionary grain generally less than 2.0 mm in diameter. In section ooids display concentric, radial or combined cortex structures.

Note: Ooid is the particle; oolite is the rock composed of ooids. See also Definition 137.

NE oöide; oölietbolletje
DE Ooide
ES oolito; ooide
FR oolithe; oolite; ooide
RE OD
AB Oo

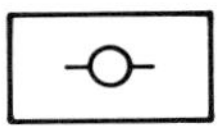

59. SUPERFICIAL OOID: a particle with a coating of one layer.

NE oppervlakkig gelaagde oöide
DE Einfachooïde; Rindenooide; Mantelooide
ES oolito superficial; ooide superficial
FR ooide a cortex réduit a une seule couche
Re OD
AB Oo, spf

60. PSEUDO-OOID: a rounded mudstone particle with firm outline, the origin of which cannot be identified with certainty.

NE schijnooöide
DE Pseudooid; Pelitkörner; Peloide

ES pseudoolito; pseudooide
FR pseudo-ooide
RE OD
AB psoo

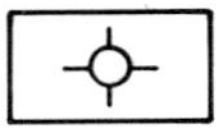

61. OOPELLET: a spherical or subspherical grain displaying characteristics of both an ooid and a pellet; the inner part is pelletoidal but it has an accretionary coating.

NE oöpellet
DE Oopellet; Einfachooide
ES pelet-oolito
FR boulette oolitique
RE PTOD
AB Pel, Oo

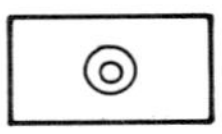

62. PISOID: a multilayered carbonate particle, reminiscent of an ooid but generally 2 mm or more in diameter, often with laminae of irregular thickness and predominantly formed in non-marine environments.

Note
1. Pisoid is the particle; a rock composed predominantly of pisoids is a pisolite. See also Definition 138.
2. Pisoids have been claimed to form in a similar manner as ooids or *in situ* in vadose diagenetic environments.

NE pisoïde
DE Pisoid
ES pisolito
FR pisolithe; pisolite
RE PD
AB Piso

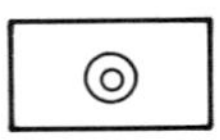 VA

63. VADOSE PISOID: a diagenetic particle commonly found in the freshwater vadose zone. Its appearance is that of a coated grain; it commonly has fitted contacts and downward elongations.

Note: See Dunham (1969b) and Esteban (1976).

NE vadose pisoïde
DE Vadose Pisoide
ES pisolito vadoso
FR pisolithe vadose; pisolite vadose
RE PD
AB vadose piso

64. ONCOID (ONKOID): growth of stacked hemispheroidal or concentric algal laminae around a free-moving flake of mud or another particle (see Table 1–1).

Note: Oncoid is the particle; oncolite is rock containing a great amount of oncoids. See also Definitions 140 and 163 and p. 59.

NE onkoide
DE Onkoid
ES oncolito
FR oncolithe; oncolite
RE OK
AB Onk, Onkd

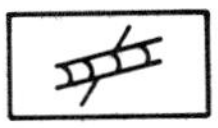

65. PSEUDO-ONCOIDS (ALGAL PELLETS; ALGAL GRAINS): globular algal bodies or nodules similar to oncoids but lacking stacked hemispheroidal or concentric laminations.

NE schijnonkoide
DE Micritonkoide
ES pseudo-oncolito
FR pseudo-oncolithe, pseudo-oncolite

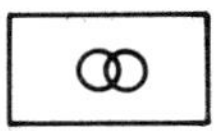

66. LUMP (BAHAMITE): a composite grain with a lobate outline composed of grains texturally similar to the material in which lumps occur, normally with micritic cements between the grains.

Note: Bahamite is an obsolete term. Its use is not recommended.

NE klont
DE Traubenklumpen; Aggregatkörner
ES lump; agregado
FR agrégat; grain composé
RE PSAG
AB Gpst

Table 1–1
Possible Genetic/Descriptive Classifications of Oncoids[a]

	Diameter nucleus/thickness coating	
Nucleus	≥ 1	< 1
Single	Superficial Superficial single	Normal Normal single
Compound	Superficial compound	Normal compound

[a] From Dahanayake (1977).

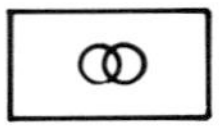

67. GRAPESTONE (BAHAMITE): a compound grain made up of various carbonates and grains, normally with sparry calcite between the grains.

Note: Bahamite is an obsolete term. Its use is not recommended.

NE samengestelde korrel
DE Grapestone; Aggregatkörner; Traubenklompen
ES grapestone
FR agrégat; grain composé
RE PSAG
AB Gpst

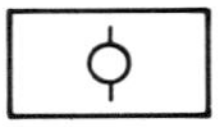

68. PELLET: a calcareous sand-sized grain composed of micritic matter and lacking distinct internal structure; generally ovoid in shape.

NE pellet; propje; bolletje
DE Pille; Peloide
ES pelet
FR boulette; pelote
RE PT
AB Pel

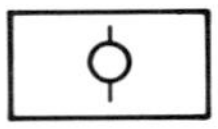

69. FAECAL PELLET: petrified excrement. It generally is produced by crustaceans and worms. It commonly is an important constituent of ancient limestone and generally has the appearance of a normal pellet. It can have structures as a result of passing through the digestive tract of organisms. A characteristic faecal pellet with digestive tract features is Favreina.

NE faecespropje; uitwerpsel propje
DE Kotpille
ES pelet fecal
FR coprolithe; boulette fécale; pelote fécale
RE PTFE
AB Pel, fec.

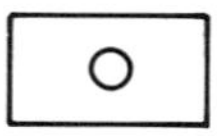

70. CLOT: a particle composed of micritic matter and lacking significant internal structure, having a fuzzy outline in clear spar. Clots are likely to be of diagenetic origin rather than sedimentary grains.

Note: See Definition 368.

NE klont
DE Pelletoide
ES peloide de contorno indefinido

FR grain micritique; pelletoïde
RE PT

71. (MICRO)PELLETOID: a micritic particle up tc sand size.

NE (mikro)pelletachtig
DE Mikropelletoide
ES (micro)peletoide
FR micropelletoïde
RE PM
AB Micrpeld

NOTE: See Definitions 68–71. The main prccesses of pellet formation are

1. Aggregation of fine particles
2. *In situ* hardening of unoriented cryptocrystalline aragonite
3. Conversion of pre-existing organic or non-organic particles

These processes are usually biological and physical. Examples of 1 are aggregation of silt and clay-sized particles by formation of faecal pellets or binding by organic slimes. Examples of 2 are internal cementation and possibly recrystallisation. Examples of 3 are biological and chemical influences leading to breakdown. Here also mechanical processes are responsible. The results of these processes are pellets, pelletoids, micropelletoids and clots (Figs. 1–5 and 1–6).

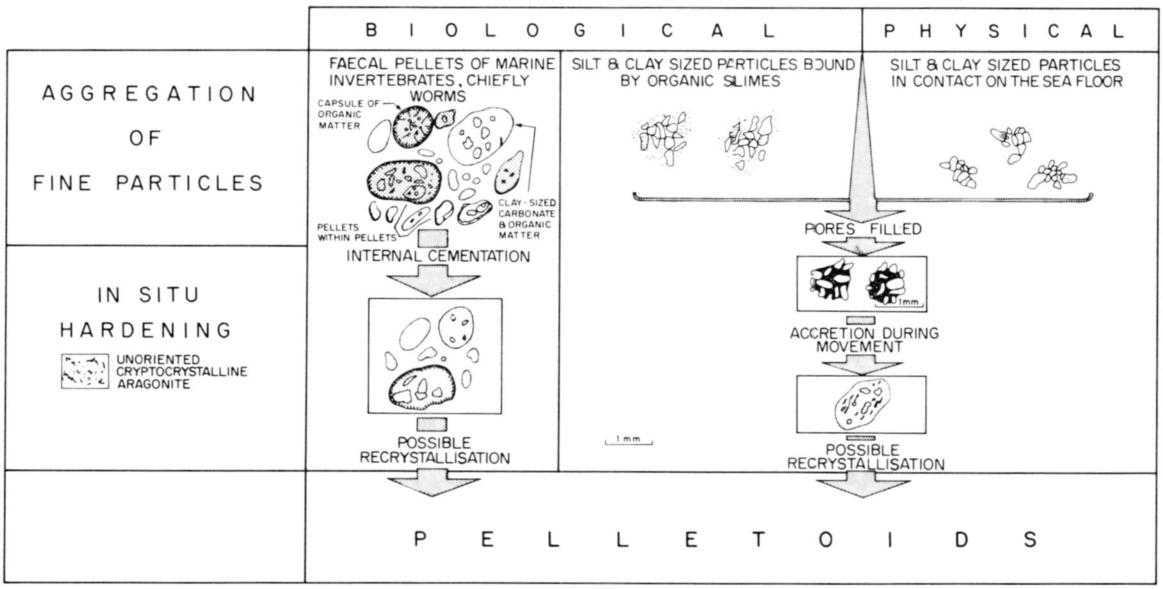

Fig. 1–5. Cementation of biological and physical aggregates and genesis of pelletoids.

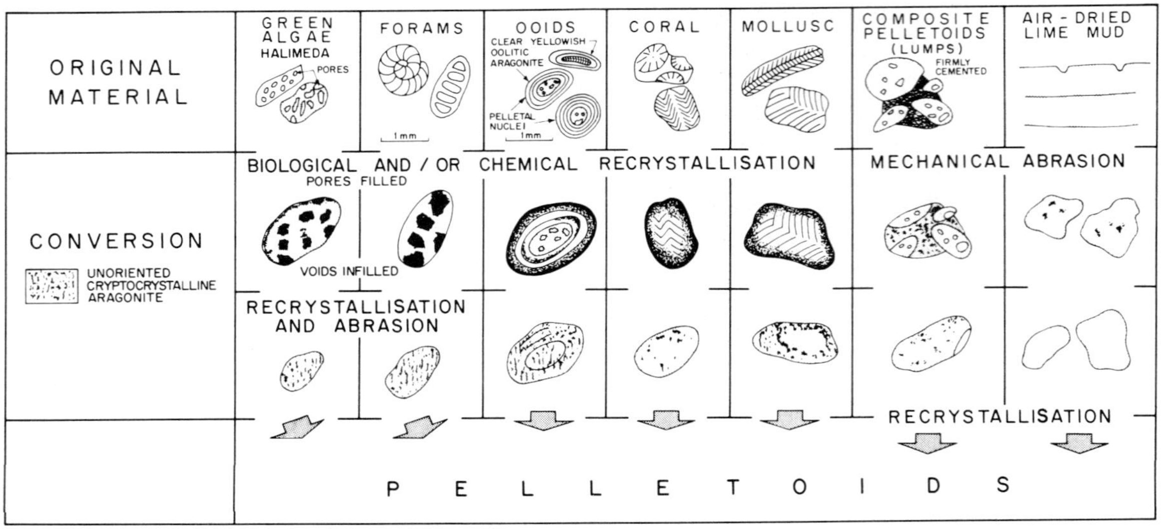

Fig. 1–6. Conversion of various particles to pelletoids.

1–4 CARBONATE DEPOSITIONAL TEXTURES AND SELECTED CLASSIFICATIONS

Introduction

Classification systems are descriptive (objective and precise) or genetic, giving information about what is known, inferred, or believed to be the origin of a rock. The best practical classifications are based on descriptive parameters into which genetic interpretations (when they can reasonably be inferred) are carefully blended.

The common basis of modern carbonate classification systems (see Ham, 1962) is the description of relative abundances of the three basic carbonate constituents: particles, interparticulate matter and pores. The various particles and the interparticulate matter have been defined and discussed in Section 1–3. Because of their complex origin, pores in carbonates are discussed in Chapters 3 and 4.

The polygenetic nature of carbonates (see the introduction to Section 1–1) inhibits easy classification. Defining parameters have to be easily observed and genetically significant, but one such set of parameters, applicable to all carbonates, is difficult to find. Grabau (1913) proposed one of the first rational schemes: a nomenclature based on the average size of the particles. His classification, to which prefixes are added denoting mineralogy, is here first defined. But grain sizes in carbonates are not always very meaningful in environmental terms. Sediment behaviour depends to a large extent on the shape and density of particles. The density of calcite (specific gravity (S.G.) 2.72), aragonite (S.G. 2.95) and quartz (S.G. 2.65) causes grains of sizes smaller than ~0.125 mm to behave similarly (Orme, 1977) but, with grains of coarse grades, shape and microstructure become increasingly important. Spheres or small blocks (e.g., forams) take a direct path of settling, but plates and rods tend to oscillate in their fall. At larger sizes, bulk density is sometimes of even greater importance than shape.

Because of the dependence of sediments on shape and density, Grabau's grain-size classification is only of limited importance in carbonates. Therefore, other classification systems are next discussed. Because of the great abundance of classification systems, the discussion is restricted to those most widely used in the oil industry.

Folk (1959) introduced a classification based on the kind and proportions of particles and the matrix ratio. The critical achievement of Folk's classification is that it forces an examination of what lies between the particles—sparry calcite cement or micrite. His terminology is self-explanatory, although occasionally rather cumbersome. Each term is a composite. The prefix indicates the nature of the particle, the stem indicates the nature of the cement or matrix and the suffix indicates the grain size (Grabau's terms). Folk's classification is purely descriptive; no genetic implications are included in his terms.

Dunham (1962) placed greater emphasis on the depositional texture of the carbonate sediments. He recognised five fundamental types based on the relative quantity of grains present (grain support or mud support) and/or the presence or absence of mud. His terminology was subsequently extended by Embry and Klovan (1972). Both the original terminology and the additions are descriptive, and genetic implications are easy to make.

Another way of looking at carbonate classification was put forward by Pettijohn (1975), who pointed out that a distinction between autochthonous and allochthonous (in the sense of intrabasinal versus extrabasinal) mechanically transported carbonates is occasionally useful. To a large degree the allochthonous carbonates can be treated as a separate group, the terminology of which can be directly compared with that of siliciclastics. Thus, the recent increased attention on deeper marine carbonates (Enos, 1983; Cook and Mullins, 1983; Scholle *et al.*, 1983; Crevello and Harris, 1985; Cook *et al.*, 1983) usually leads to descriptions employing terminology that is equally valid in descriptions of siliciclastics. It has to be remembered, however, that the easy solubility of carbonate minerals and their low mechanical resistance usually precludes large transport distances; therefore, extrabasinal carbonates are rare. Such allochthonous carbonates exhibit a particle–interparticle relationship analogous to that found in siliciclastic sandstones.

Autochthonous carbonates encompass such units as bioherms, biostromes, mud mounds and algal–trapped carbonates amongst many other types. The subdivision in the present section is as follows:

Grabau's: Grain-size terms
Folk's: Grain-size, grain-kind, minerals terms
Dunham's: Grain-abundance, mud-presence terms (with additions by Embry and Klovan)
Monogranulate carbonate terms
Hybrid terms

Definitions 72–88 (Grabau's Grain-Size Terms)

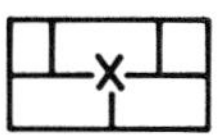

72. CALCIMICRITE: a limestone, the particles and crystals of which have diameters of less than 0.004 mm.

Note: The cutoff value (0.004 mm) between mud-sized particles and others used by Folk (1959) is different from that (0.02 mm) used by Dunham (1962).

NE calcimikriet
DE Kalkmikrit

ES calcimicrita
FR micrite; calci-micrite
RE PILS

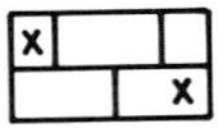

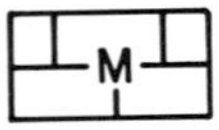

73. CALCILUTITE (CALCIPELITE): a limestone consisting entirely of calcite particles up to 0.06 mm. These are both original sediment particles and crystals as a result of diagenetic alteration of silt and/or clay size. The term is also used in a wider sense, to include chemically precipitated crystalline sedimentary components.

NE kalkslibsteen; calcilutiet
DE Kalklutit; Kalkpelit
ES calcilutita; calcipelita
FR calcilutite; pétite calcaire
RE PILS; LTLS

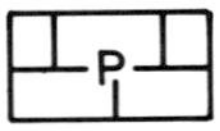

74. CALCISILTITE: a limestone consisting predominantly of (detrital) carbonate particles of silt size.

NE kalkslibsteen
DE Kalksiltit
ES calcisiltita
FR aleurolite calcaire; calcisiltite
RE SJLS

GRAIN SIZE (Wentworth's scale, slightly modified)

mm	μm	φ	visual		NOMENCLATURE	Abbreviations informal	Abbreviations computer
				Rudite	Boulder	Bld	BO
256		-8					
				Rudite	Cobble	Cbl	CB
64		-6					
				Rudite	Pebble	Pbl	PB
4		-2					
				Rudite	Granule	Gran	GR
2		-1					
				Arenite	Very coarse	crs	R
1		0					
				Arenite	Coarse	crs	C
1/2	500	1					
				Arenite	Medium	m	M
1/4	250	2					
				Arenite	Fine	f	F
1/8	125	3					
				Arenite	Very fine	f	V
1/16	63	4					
				Lutite	Silt	Slt	SJ
1/50	20	5.65					
				Lutite	Pelite	Pel	PI

Fig. 1–7. Wentworth's grain-size scale, slightly modified as used in Shell's standard legend. For practical reasons Wentworth's division of the lutites into clay and silt at the 4-μm (1/256 mm) boundary has been replaced by the subdivision into pelite and silt at the 20-μm boundary, as shown here. The parameter ϕ is $-\log_2$ diameter in millimeters.

P,G

75. CALCARENITE: a limestone consisting predominantly of (detrital) calcite particles of sand size; a consolidated calcareous sand.

NE kalkzandsteen
DE Kalkarenit
ES calcarenit
FR calcarénite
RE ARLS
AB Sst, carb.

	Transported Constituents	Authigenic Constituents	
	Very coarse calcirudite	Extremely coarsely crystalline	
64 mm			
	Coarse calcirudite		
16 mm			
	Medium calcirudite		
4 mm			4 mm
	Fine calcirudite	Very coarsely crystalline	
1 mm			1 mm
	Coarse calcarenite	Coarsely crystalline	
0.5 mm			
	Medium calcarenite		
0.25 mm			0.25 mm
	Fine calcarenite	Medium crystalline	
0.125 mm			
	Very fine calcarenite		
0.062 mm			0.062 mm
	Coarse calcilutite	Finely crystalline	
0.031 mm			
	Medium calcilutite		
0.016 mm			0.016 mm
	Fine calcilutite	Very finely crystalline	
0.008 mm			
0.004 mm	Very fine calcilutite		0.004 mm
		Aphanocrystalline	

Fig. 1–8. Grain size scale for allochthonous and autochthonous carbonate sediments. Carbonate rocks contain both physically transported particles (oolites, intraclasts, fossils and pellets) and chemically precipitated minerals (as pore-filling cement, primary ooze or products of recrystallisation and replacement). Therefore, the size scale must be a double one, so that one can distinguish which constituent is being considered (e.g., calcirudites may be cemented with very fine crystalline dolomite, and finer calcarenites may be cemented with coarsely crystalline calcite).

The size scale for transported constituents uses the terms of Grabau but retains the finer divisions of Wentworth except in the calcirudite range; for dolomites of obviously allochemical origin, the terms *dolorudite, doloarenite* and *dololutite* are substituted for those shown. The most common crystal size for dolomite appears to be between 0.062 and 0.25 mm, and for this reason that interval was chosen as the *medium crystalline* class (from Folk, 1962).

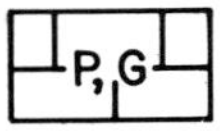

76. CALCARENITIC LIMESTONE: a limestone composed of more than 10% of original calcareous mud matrix (particles with diameters of less than 0.04 mm) accompanied by more than 10% coarser carbonate grains.

NE kalkzandsteen-achtige kalk
DE kalkarenitischer Kalkstein
ES caliza calcarenítica
FR calcaire calcarénitique; calcarénite
RE LS AR

77. CALCARENACEOUS: said of any sandstone containing a significant amount (up to 50%) of calcium carbonate detritus (see Fig. 1–9).

NE kalkzandig
DE kalkdetritusführender Sandstein
ES calcarenítica
FR grés calcarénitique; grés à éléments calcaires
RE LS
AB calc

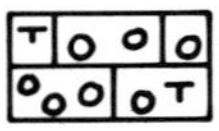

78. CALCIRUDITE: a limestone consisting predominantly (more than 50%) of detrital calcite particles larger than 2 mm in diameter and often also cemented with calcareous material.

NE kalkkiezelsteen; calcirudiet
DE Kalkrudit

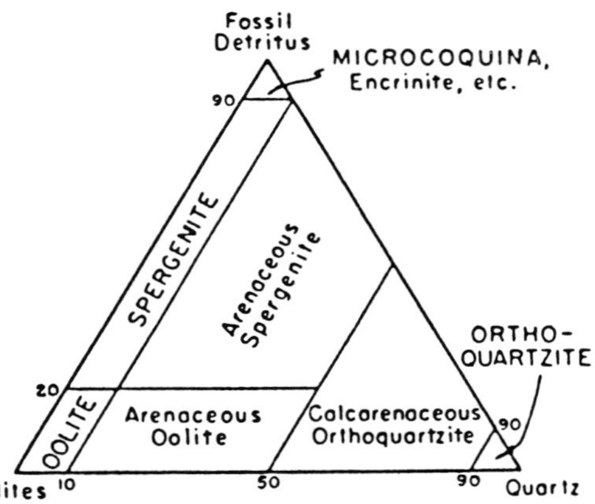

Fig. 1–9. Classification of the calcarenites and related rocks based on the composition of the framework fraction. From Pettijohn (1975) Fig 10–34 (p. 353). Copyright 1949 © 1957 by Harper & Row, Publishers, Inc. Copyright © 1975 by Francis J. Pettijohn. Reprinted by permission of the publisher.

ES calcirudita
FR calcirudite
RE RULS
AB Rud, calc

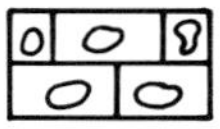

79. CALCIGRAVEL: unconsolidated equivalent of calcirudite.

NE kalkkiezel
DE Kalkgerölle
ES grava calcárea
FR gravier calcaire
RE GVLS
AB Grv, calc.

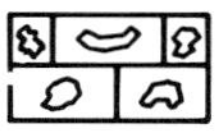

80. CALCIBRECCIA: calcirudite with angular lithoclasts.

NE kalkbreccie
DE Kalkbrekzie; Kalkbreccie
ES calcibrecha; brecha calcárea
FR brèche calcaire
RE BRLS
AB Brc, calc

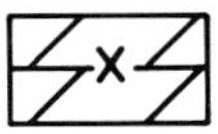

81. DOLOMICRITE (DOLOMITE MUDSTONE): a carbonate sedimentary rock containing less than 1% allochems, interpreted as a lithified dolomite mud and made up of mud-sized (<0.04 mm) dolomite crystals.

Note: See also Definitions 9 and 83.

NE kristallijne dolomiet slibsteen; dolomicriet
DE Dolomikrit
ES dolomicrita
FR dolomicrite; boue dolomitique
RE PIDM
AB Mcr, dol

82. DOLOSPARITE: a sparry dolomitic crystalline carbonate.

Note: See also definition 9.

NE dolospariet
DE Dolosparit
ES doloesparita
FR dolosparite
RE SP DM

83. DOLOLUTITE: (1) a dolomitic rock consisting predominantly of dolomite particles or crystals of silt and/or mud size (<0.04 mm); (2) a consolidated dolomitic mud.

Note: See also Definitions 9 and 81.

NE dolomitische slibsteen
DE Dololutit
ES dololutita
FR dololutite
RE PIDM
AB Lut, dol.

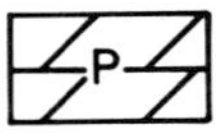

84. DOLOSILTITE: a dolomite rock consisting predominantly of detrital dolomite particles of silt size; a consolidated dolomitic silt.

NE dolomitische slibsteen
DE Dolosiltit
ES dolosiltita
FR aleurolite dolomitique; dolosiltite
RE SIDM
AB Slt, dol

85. DOLARENITE: (1) a dolomite rock consisting predominantly of dolomite particles or crystals of sand size; (2) a consolidated dolomitised detrital limestone.

NE dolomiet zandsteen
DE Dolarenit
ES doloarenita
FR dolarénite
RE AR DM
AB Sst, dol

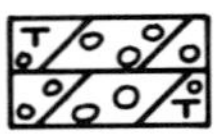

86. DOLORUDITE: a dolomite rock consisting predominantly of detrital dolomite particles larger than sand size; a consolidated dolomitic gravel.

NE dolomitische kiezelsteen
DE Dolorudit
ES dolorudita
FR dolorudite
RE RU DM
AB Rud, dol

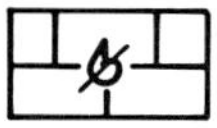

87. BIOCLASTIC LIMESTONE: carbonate rock composed of cemented skeletons (or skeletal fragments).

NE bioklastisch gesteente
DE bioklastisches Gestein
ES roca bioclástica
FR calcaire bioclastique
RE RC SF

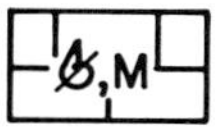

88. BIOCALCILUTITE: a calcilutite containing bioclasts.

NE kalkslibsteen met skeletfragmenten
DE Biokalklutit
ES biocalcilutita
FR biocalcilutite
RE PISF; LTSF

Definitions 89–122 (Folk's Grain-Size, Grain-Kind and Mineral Terms; see Figs. 1–10 to 1–12)

PREFIXES: "Bio" is used with a rock term to indicate that the rock contains a significant, usually abundant quantity of sedimentary particles of skeletal origin. "Litho", "intra", "pel" and "oo" are used with a rock term to indicate that the rock contains a significant, usually abundant quantity of lithoclasts, intraclasts, pellets or ooids, respectively. "Dolo" is used with a rock term to indicate that the rock mineralogy is dolomite.

89. MICRITE (**micr**ocrystalline ca**lcite**): (1) an inorganic and/or biochemical precipitate, formed within the environment of deposition and showing little or no evidence of significant transport; (2) a microcrystalline limestone made up almost entirely of crystalline calcite; (3) a microcrystalline limestone with 1 to 10% allochems, in which case a prefix should be added, for example, ooid-bearing, fossiliferous (oo-, bio-); (4) matrixes of microcrystalline calcitic textural composition. Crystal sizes do not exceed 4 μm (0.004 mm) (Folk, 1959); 31 μm (0.031 mm) (Leighton and Pendexter, 1962).

Note
1. It should be realised that micrite (*sensu* Folk) is not the same as lime mud (*sensu* Dunham). Folk gives a descriptive term for a particular rock; Dunham interprets a particular sediment.
2. See also Definition 16.
3. For sparite see Definition 341.

NE micriet
DE Mikrit

Volumetric Allochem Composition	Limestones, Partly Dolomitised Limestones, and Primary Dolomites (see Notes 1 to 6)					Replacement Dolomites[7] (V)	
	>10% Allochems Allochemical Rocks (I and II)		<10% Allochems Microcrystalline Rocks (III)				
	Sparry Calcite Cement > Microcrystalline Ooze Matrix	Microcrystalline Ooze Matrix > Sparry Calcite Cement	1–10% Allochems	<1% Allochems	Undisturbed Bioherm Rocks (IV)	Allochem Ghosts	No Allochem Ghosts
	Sparry Allochemical Rocks (I)	Microcrystalline Alochemical Rocks (II)	Most Abundant Allochem			Evident Allochem	
>25% Intraclasts (i)	Intrasparrudite (Ii:Lr) Intrasparite (Ii:La)	Intramicrudite* (IIi:Lr) Intramicrite* (IIi:La)	Intraclasts: Intraclast-bearing Micrite* (IIIi:Lr or La)	Micrite (IIIm:L); if disturbed, Dismicrite (IIImX:L); if primary dolomite, Dolomicrite (IIIm:D)	Biolithite (IV:L)	Finely Crystalline Intraclastic Dolomite (Vi:D3) etc.	Medium Crystalline Dolomite (V:D4)
<25% Intraclasts; >25% Oolites (o)	Oosparrudite (Io:Lr) Oosparite (Io:La)	Oomicrudite* (IIo:Lr) Oomicrite* (IIo:La)	Oolites: Oolite-bearing Micrite* (IIIo:Lr or La)			Coarsely Crystalline Oolitic Dolomite (Vo:D5) etc.	Finely Crystalline Dolomite (V:D3)
<25% Intraclasts; <25% Oolites; Volume Ratio of Fossils to Pellets >3:1 (b)	Biosparrudite (Ib:Lr) Biosparite (Ib:La)	Biomicrudite (IIb:Lr) Biomicrite (IIb:La)	Fossils: Fossiliferous Micrite (IIIb: Lr, La, or Ll)			Aphanocrystalline Biogenic Dolomite (Vb:Dl) etc.	
<25% Intraclasts; <25% Oolites; Volume Ratio of Fossils to Pellets 3:1–1:3 (bp)	Biopelsparite (Ibp:La)	Biopelmicrite (IIbp:La)	Pellets: Pelletiferous Micrite (IIIp:La)			Very Finely Crystalline Pellet Dolomite (Vp:D2) etc.	etc.
<25% Intraclasts; <25% Oolites; Volume Ratio of Fossils to Pellets <1:3 (p)	Pelsparite (Ip:La)	Pelmicrite (IIp:La)					

Fig. 1–10. Classification of carbonate rocks according to Folk (1962). By permission of the AAPG. The notes to the classification scheme are as follows:

* Designates rare rock types.

[1] Names and symbols in the body of the table refer to limestones. If the rock contains more than 10% replacement dolomite, prefix the term *dolomitised* to the rock name, and use DLr or DLa for the symbol (e.g., dolomitised intrasparite, Li : DLa). If the rock contains more than 10% dolomite of uncertain origin, prefix the term *dolomitic* to the rock name, and use dLr or dLa for the symbol (e.g., dolomitic pelsparite, Ip : dLa). If the rock consists of primary (directly deposited) dolomite, prefix the term *primary dolomite* to the rock name, and use Dr or Da for the symbol (e.g., primary dolomite intramicrite, IIi : Da). Instead of *primary dolomite micrite* (IIIm : D) the term *dolomicrite* may be used.

[2] Upper name in each box refers to calcirudites (median allochem size larger than 1.0 mm); lower name refers to all rocks with median allochem size smaller than 1.0 mm. Grain size and quantity of ooze matrix, cements or terrigenous grains are ignored.

[3] If the rock contains more than 10% terrigenous material, prefix sandy, silty or clayey to the rock name, and Ts, Tz or Tc to the symbol, depending on which is dominant (e.g., sandy biosparite, TsIb : La, or silty dolomitized pelmicrite, TzIIp : DLa). Glauconite, collophane, chert, pyrite or other modifiers may also be prefixed.

[4] If the rock contains other allochems in significant quantities that are not mentioned in the main rock name, these should be prefixed as qualifiers preceding the main rock name (e.g., fossiliferous intrasparite, oolitic pelmicrite, pelletiferous oosparite or intraclastic biomicrudite). This can be shown symbolically as Ii(b), Io(p), IIb(i), respectively.

[5] If the fossils are of rather uniform type or one type is dominant, this fact should be shown in the rock name (e.g., pelecypod biosparrudite, crinoid biomicrite).

[6] If the rock was originally microcrystalline and can be shown to have recrystallized to microspar (5–15 μm, clear calcite) the terms *microsparite, biomicrosparite*, etc., can be used instead of *micrite* or *biomicrite*.

[7] Specify crystal size as shown in the examples.

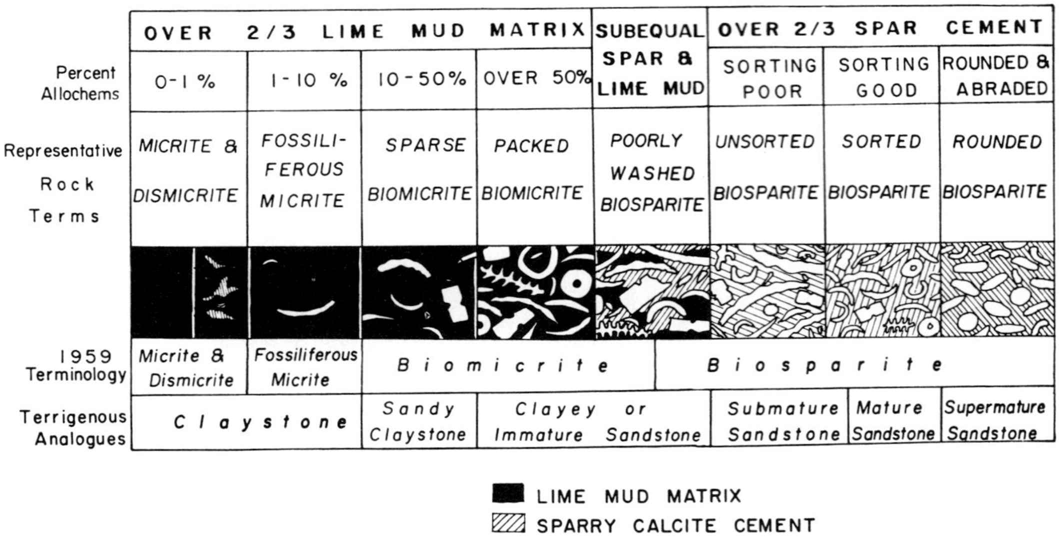

Fig. 1–11. Classification of carbonate textures according to Folk (1962). By permission of the AAPG.

ES micrita
FR vase micritique; micrite; calcaire microcristallin; calcaire micritique; boue micritique

90. ORTHOMICRITE: micrite formed by direct chemical precipitation.

NE primaire mikriet
DE Orthomikrit
ES ortomicrita
FR orthomicrite

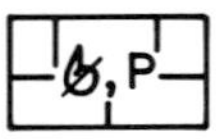

91. BIOMICRITE: a carbonate rock consisting predominantly of skeletal fragments and micrite. Other components such as intraclasts and ooids occur in quantities smaller than 25%. The major organism is usually specified when the term is used (e.g., crinoid biomicrite).

NE biomikriet
DE Biomikrit
ES biomicrita
FR biomicrite
RE MX SF

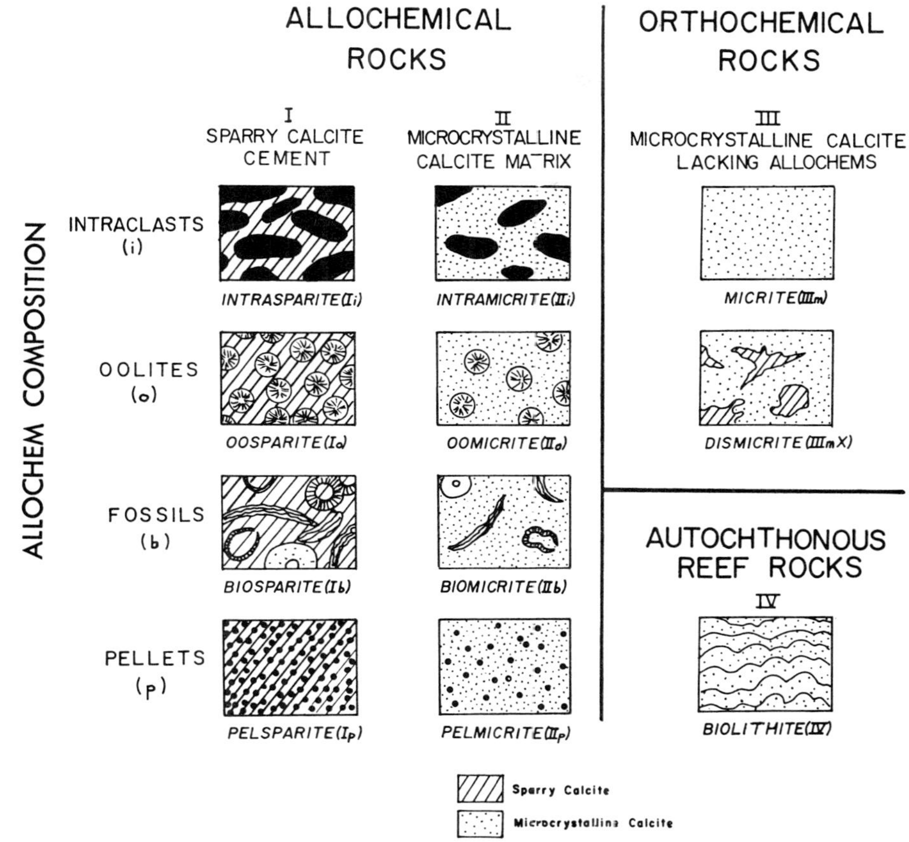

Fig. 1–12. Classification of allochemical and orthochemical rocks according to Folk (1962). By permission of the AAPG.

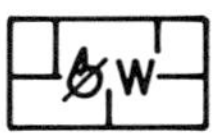

92. SPARSE BIOMICRITE: a biomicrite in which bicclasts make up 10 to 50% of the rock.

NE bioklast-arme biomikriet
DE locker gepackter Biomikrit
ES sparse biomicrita
FR micrite à bioclastes épars
RE MX SF

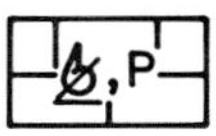

93. PACKED BIOMICRITE: a biomicrite in which the bioclasts make up more than 50% of the rock.

NE bioklast-rijke biomikriet
DE dicht gepackter Biomikrit
ES packed biomicrita
FR micrite à bioclastes abondants
RE MXSF

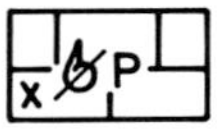

94. BIOMICROSPARITE: a biomicrite in which the carbonate mud matrix has recrystallised to microspar; also a microsparite containing fossils or fossil fragments.

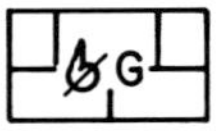

NE biomikrospariet
DE Biomikrosparit
ES biomicroesparita
FR biomicrosparite
RE SPMX

95. BIOMICRUDITE: a biomicrite containing fossils or fossil fragments that are more than 2 mm in diameter.

NE biomikrudiet
DE Biomikrudit
ES biomicrudita
FR biomicrudite
RE MXSF

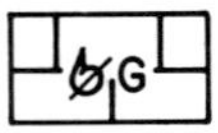

96. BIOSPARITE: a carbonate rock consisting predominantly of skeletal fragments and sparite. Other components such as intraclasts and ooids occur in quantities smaller than 25%. Sorting of bioclasts allows further textural subdivision. The major organism is usually specified when the term is used (e.g., pelecypod biosparite).

NE biospariet
DE Biosparit
ES bioesparita
FR biosparite
RE SPSF

97. BIOSPARARENITE: a biosparite containing sand-sized bioclasts.

NE biosparzandkalksteen
DE Biospararenit
ES bioesparárenita

FR biosepararénite
RE SPSF

98. BIOSPARRUDITE: a biosparite containing bioclasts that are more than 2 mm in diameter.

NE biosparudiet
DE Biosparrudit
ES bioesparrudita
FR biosparrudite
RE SPSF

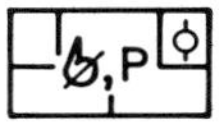

99. BIOPELMICRITE: a carbonate containing less than 25% intraclasts and less than 25% ooids with a volume ratio of fossils to pellets ranging from 3 : 1 to 1 : 3 and the carbonate mud matrix (micrite) more abundant than the sparry calcite cement.

NE biopelmikriet
DE Biopelmikrit
ES biopelmicrita
FR biopelmicrite
RE MXPT

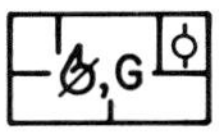

100. BIOPELSPARITE: a carbonate containing less than 25% intraclasts and less than 25% ooids with a volume ratio of fossils to pellets ranging from 3 : 1 to 1 : 3 and the sparry calcite cement more abundant than the carbonate mud matrix (micrite).

NE biopelspariet
DE Biopelsparit
ES biopelesparita
FR biopelsparite
RE SPPT

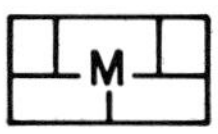 L

101. LITHOCALCILUTITE: a calcilutite containing abundant lithoclasts.

NE detritische kalkslibsteen
DE Lithokalklutit
ES litocalcilutita
FR lithocalcilutite
RE LTCL
AB Lut, lithocalc.

102. LITHODOLOLUTITE: a dololutite containing abundant lithoclasts.

NE detritische dolomitische slibsteen
DE Lithodololutit
ES litodololutita
FR lithodololutite
RE PICL

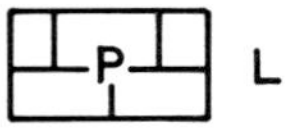

103. LITHOCALCISILTITE: a calcisiltite containing abundant lithoclasts.

NE detritische kalkslibsteen
DE Lithokalksiltit
ES litocalcisiltita
FR lithocalcisiltite
RE SJCL

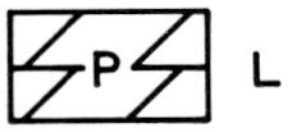

104. LITHODOLOSILTITE: a dolosiltite containing abundant lithoclasts.

NE detritische dolomitische slibsteen
DE Lithodolosiltit
ES litodolosiltita
FR lithodolosiltite
RE SJCL

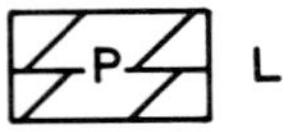

105. LITHOCALCARENITE: a calcarenite containing abundant lithoclasts.

NE detritische kalkzandsteen
DE Lithokalkarenit
ES litocalcarenita
FR lithocalcarénite
RE ARCL

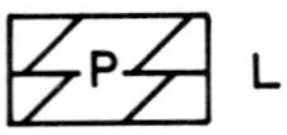

106. LITHODOLARENITE: a dolarenite containing abundant lithoclasts.

NE detritische dolomitische zandsteen
DE Lithodoloarenit
ES litodolarenita
FR lithodolarénite
RE ARCL

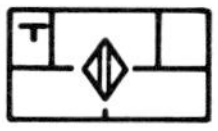

L

107. LITHOCALCIRUDITE: a calcirudite containing abundant lithoclasts.

NE detritische kalkkiezelsteen
DE Lithokalkrudit
ES litocalcirudita
FR lithocalcirudite
RE RULC

L

108. LITHODOLORUDITE: a dolorudite containing abundant lithoclasts.

NE detritische dolomitische kiezelsteen
DE Lithodolorudit
ES litodolorudita
FR lithodolorudite
RE RULC

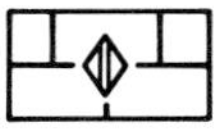

I

109. INTRAMICRITE: a carbonate with more than 10% allochems, more than 25% of which are intraclasts in a micritic matrix.

NE intramikriet
DE Intramikrit
ES intramicrita
FR intramicrite

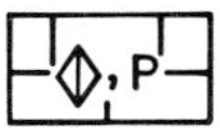

I

110. INTRAMICARENITE: an intramicrite containing sand-sized intraclasts.

NE intramikareniet
DE Intramikarenit
ES intramicarenita
FR intramicarénite

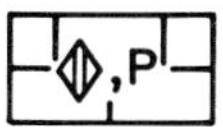

I

111. INTRAMICRUDITE: an intramicrite containing gravel-sized intraclasts.

NE intramikrudiet
DE Intramikrudit
ES intramicrudita
FR intramicrudite

I

112. INTRASPARITE: carbonate with more than 10% allochems, more than 25% of which are intraclasts and in which the sparry calcite cement is more abundant than the carbonate mud matrix (micrite). The spar is usually intergranular cement.

NE intraspariet
DE Intrasparit
ES intraesparita
FR intrasparite

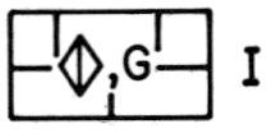

113. INTRASPARARENITE: an intrasparite containing sand-sized intraclasts.

NE intrasparzandkalksteen
DE Intrasparerenit
ES intraesparrarenita
FR intrasparrarénite

114. INTRASPARRUDITE: an intrasparite containing gravel-sized intraclasts.

NE intrasparrudiet
DE Intrasparrudit
ES intraesparrudita
FR intrasparrudite

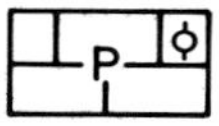

115. PELLETAL LIMESTONE: limestone consisting predominantly of pellets.

NE kalksteen van pellets
DE peloidführender Kalk(stein)
ES caliza peletífera
FR calcaire à pelletoïdes
RE LS PT
AB Lst, pel

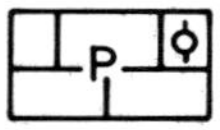

116. PELMICRITE: a carbonate containing less than 25% intraclasts and less than 25% ooids with a volume ratio of pellets to bioclasts of greater than 3 : 1 and the carbonate mud matrix more abundant than the sparry calcite cement.

NE pelmikriet
DE Pelmikrit
ES pelmicrita
FR pelmicrite
AB Mcr, pel

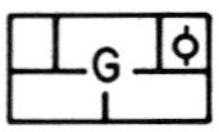

117. PELSPARITE: a carbonate containing less than 25% intraclasts and less than 25% ooids with a volume ratio of pellets to bioclasts of greater than 3 : 1 and the sparry calcite cement more abundant than the carbonate mud matrix (micrite).

NE pelspariet
DE Pelsparit
ES pelesparita
FR pelsparite
AB Spar, pel

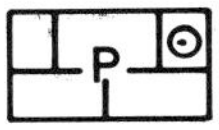

118. OOMICRITE: a carbonate containing at least 25% ooids and no more than 25% intraclasts and in which the carbonate mud matrix (micrite) is more abundant than the sparry calcite cement.

NE oömikriet
DE Oomikrit
ES oomicrita
FR oomicrite
AB Mcr, Oo

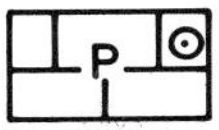

119. OOMICRUDITE: an oomicrite containing ooids that are more than 2 mm in diameter.

Note: Strictly speaking such a rock should be referred to as a pisomicrudite. (See Definition 50.)

NE oömikrudiet
DE Oomikrudit
ES oomicrudita
FR oomicrudite

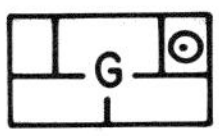

120. OOSPARITE: a carbonate containing at least 25% ooids and no more than 25% intraclasts and in which the sparry calcite cement is more abundant than the carbonate mud matrix (micrite). The spar is usually an intergranular cement.

NE oöspariet
DE Oosparit
ES ooesparita
FR oosparite
AB Spar, oo

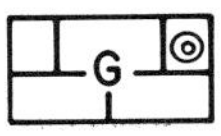

121. OOSPARUDITE: an oosparite containing ooids that are more than 2 mm in diameter.

Note: Strictly, speaking such a rock type should be referred to as a pisosparudite. (See Definition 50.)

NE oösparudiet
DE Oosparrudit

ES ooesparrudita
FR oosparudite

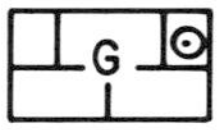

122. OOSPAR(AR)ENITE: an oosparite containing medium-coarse and sand-sized particles; an oolitic sandstone.

NE oölitische zandalksteen
DE Oosparararenit
ES ooespar(ar)enita
FR oospararénite

Dunham's Grain-Abundance, Mud-Presence Terms (Fig. 1–13) with Embry–Klovan Additions (Fig. 1–14)

Depositional Textures

Depositional textures result from the spatial arrangement and the relative abundances of the various constituents of sedimentary rocks. In classifications based on physically determined textural character such as Graham's, the origin of particles is insignificant, and carbonates are subdivided according to particle size and shape. Calcirudite is thus distinguished from calcarenite. This assumes transportation of grains controlled by hydraulic conditions in the environment. Biogenic textures, however, are controlled by biological or biochemical processes responsible for producing carbonate.

As has been pointed out before (see the introduction to Section 1–4), particle size is usually insignificant and subdivision according to origin of particles is more meaningful. Crinoidal limestone is thus distinguished from coral limestone. Most carbonates are influenced by both hydraulic conditions and biological or biochemical processes in the depositional environment. These then jointly control the carbonate depositional texture. Consequently classifications taking both aspects into consideration have environmental significance. This is the case with Dunham's classification of which a few aspects will be described below.

Basis of the Dunham Classification

Role of Mud. In order to distinguish sediment deposited in calm water from that deposited in agitated water, attention could be focussed on average or predominant grain size, but one would erroneously be assuming that all particles of a specific size in a sample have the same hydraulic properties. A better textural parameter for water energy is the presence or absence of fines. Then muddy rocks and mud-free rocks are contrasted, regardless of the nature and size of included coarse material.

Role of Grains. The next useful textural parameter is the relative abundance of grains. Grains could be prominantly present (e.g., surpassing 10%) but not yet abundant enough to support one another. In such a texture the grains are "floating" in a finer-grained groundmass. The texture is called *mud supported.* Grains could also be abundant enough to support one another. Then the texture is *grain supported* (see Fig. 1–13).

An addition to this textural subdivision has been introduced by Embry and Klovan (1972) (Fig. 1–14). In specific cases (e.g., in reefoid provinces) the size, shape and nature of the carbonate

Texture				Name	Abbreviation: informal	Abbreviation: computer
Depositional texture recognisable	Original components bound together during deposition			Lime – Boundstone	Bdst	B
Depositional texture recognisable	Original components not bound together during deposition	Grain suppor-ted	Lacks mud	Lime – Grainstone	Grst	G
Depositional texture recognisable	Original components not bound together during deposition	Grain suppor-ted	Contains mud	Lime – Packstone	Pkst	P
Depositional texture recognisable	Original components not bound together during deposition	Mud supported	Particles ($>20\mu m$) $>10\%$	Lime – Wackestone	Wkst	W
Depositional texture recognisable	Original components not bound together during deposition	Mud supported	Particles ($>20\mu m$) $<10\%$	Lime – Mudstone	Mdst	M
	Original components not bound together during deposition	Mud supported	Particles ($>20\mu m$) $<10\%$	Aphanitic	aph	A
Depositional texture not recognisable	Recrystallised texture		Fine	Crystalline	xln	X
Depositional texture not recognisable	Recrystallised texture		Coarse	Sucrosic	suc	S

Fig. 1–13. Textural classification of limestones based on the classification of Dunham (1962). Slightly modified illustration from Shell's standard legend.

grains may play a role in establishing the function of the organisms that produced these grains. Differentiating mud-supported and grain-supported rocks is subjective because of the need to envision a three-dimensional arrangement of irregular shapes from the two-dimensional view offered by the microscope. Experience aids in making the right decision.

Role of Binding. Most carbonate rocks retain their depositional texture after lithification. Usually the fact that the unconsolidated sediment was composed of recognisable, discrete particles remains visible. Some carbonates, however, may show signs of having been bound during deposition. Amongst the indications for this (Dunham, 1962) are interconnected skeletal matter,

Allochthonous limestones: original components not organically bound during deposition						Autochthonous limestones: original components organically bound during deposition		
Less than 10% > 2mm components				Greater than 10% > 2mm components		By organisms which act as baffles	By organisms which encrust and bind	By organisms which build a rigid framework
Contains lime mud (<0.03 mm)			No lime mud					
Mud supported		Grain supported		Matrix supported	> 2mm component supported			
Less than 10% grains (>03mm <2mm)	Greater than 10% grains							
Mud-stone	Wacke-stone	Pack-stone	Grain-stone	Float-stone	Rud-stone	Baffle-stone	Bind-stone	Frame-stone

Fig. 1–14. Amplification of the original Dunham (1962) classification of limestones according to depositional texture by Embry and Klovan (1972) (Fig. 2). By permission of the Can. Soc. Pet. Geol.

lamination contrary to gravity and large sediment-floored cavities roofed by (questionably) organic matter. Embry and Klovan (1972) added to this organisms which act as baffles, which encrust and bind and which build a rigid framework.

Objective of Studying Carbonate Textures. Carbonate depositional textures are studied for the following reasons:

1. Together with the types of carbonate grains, they provide criteria by which hydraulic conditions and organic productivity at the time of deposition can be interpreted.
2. They have different initial permeability. This governs the rate of flow of subsurface waters and produces differences in the degree of subsequent alteration.
3. The nature and degree of subsequent alteration (i.e., diagenesis; see Chapter 3) can often be directly related to depositional textures (see Sections 3–1 and 3–2).

Textural Classification of Carbonates versus Siliciclastics. The various classes of carbonates in Dunham's classification system (1962) are distinguished primarily by different mud content. They are adequate to describe carbonate depositional textures. Unlike the situation with siliciclastics, size and sorting of grains are not significant parameters. This difference between carbonates and siliciclastics gives rise to the following observations:

1. Size and sorting of siliciclastics provide a criterion for interpreting transportation, deposition and textural maturity of a deposit. In contrast, size and sorting of carbonate grains depend on both hydraulic conditions and organic productivity at the site of deposition. The nature and abundance of produced carbonate particles determine to a large extent size and sorting of the carbonate sediment. Detailed size–sorting classifications of carbonate sediments do not necessarily lead to a better understanding of them.

2. Size and sorting of unconsolidated siliciclastics can be directly related to porosity, permeability and pore size. Carbonate grains have a wide variety of shapes, making it difficult to relate size and sorting with porosity or permeability of unconsolidated carbonate clastics. Moreover, carbonate particles frequently have depositional intragranular porosity.

3. Even after siliciclastics have been considerably modified by cementation or compaction, some direct relation of porosity and permeability to depositional texture remains. In carbonates the degree of diagenetic modification is usually more intensive, and no direct relation to size and sorting can be found. A relation expressed by the *carbonate fabric unit* concept (Reijers and Bartok, 1985); see also Definition 300) is then called for.

Definitions 123–136 (Dunham's Grain-Abundance, Mud-Presence Terms with Embry–Klovan Additions)

123. DEPOSITIONAL TEXTURE: the size, shape and spatial arrangement of grains (>0.02 mm *sensu* Dunham, 1962) and mud (<0.02 mm *sensu* Dunham, 1962) in a sediment prior to superposition of significant diagenetic changes

NE afzettings maaksel; afzettings textuur
DE Ablagerungsgefüge
ES textura deposicional

FR texture sédimentaire; texture initiale; texture aquise du moment du dépôt
AB text, dep

124. MUDDY CARBONATE ROCK: a term implying that original sediment was dominantly composed of mud-sized carbonate particles (usually a mud-supported texture). To prevent confusion with a rock containing some non-carbonate mud, a qualifier, such as ''clayey'' or ''argillaceous'', should be used for the latter case.

Note
1. This term is often used in an informal or coloquial sense.
2. See also Definitions 16 and 89.

NE karbonaatgesteente van kalkslib
DE Schlammkalk
ES caliza de grano fino
FR calcaire fin; calcaire micritique

125. MUD-SUPPORTED: term used in Dunham's classification for muddy carbonate rocks which contain particles not occurring in sufficient amounts to support one another.

NE slibondersteuning
DE schlammgestützt
ES mud-supported
FR texture flottante

126. GRAIN-SUPPORTED: term used in Dunham's classification for carbonate sedimentary rock in which sedimentary grains touch each other and form the framework of the rock.

NE korrelondersteuning
DE komponentengestützt
ES ''grain-supported''
FR texture jointive

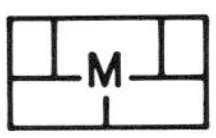

127. LIME-MUDSTONE: carbonate rock interpreted to have been originally a sediment made up of ''lime mud'' (particles smaller than 0.02 mm in diameter) with less than 10% carbonate ''grains'' (particules larger than 0.02 mm in diameter).

Note: See also Definition 6.

NE kalkslibsteen
DE Kalk-''Mudstone''; feinkörniger Kalk

ES caliza-“mudstone”
FR calcaire à grain fin; calcaire micritique
RE M
AB Mdst

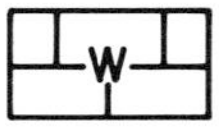

128. LIME-WACKESTONE: lime mud–supported carbonate rock containing more than 10% grains; prefix *lime* is recommended.

NE kalk-wackestone
DE Kalk-“Wackestone”
ES caliza-“wackestone”
FR calcaire à texture “wackestone”
RE W
AB Wkst

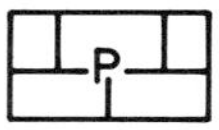

129. (LIME)-PACKSTONE: grain-supported carbonate rock with a matrix of lime mud; prefix *lime* is recommended.

NE kalk-packstone
DE Kalk-Packstone
ES caliza-packstone
FR calcaire à texture “packstone”
RE P
AB Pkst

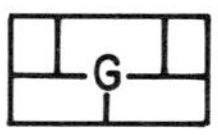

130. (LIME)-GRAINSTONE: lime mud–free carbonate rock which is grain supported.

NE kalk-“grainstone”; korrelsteen
DE Kalk-“Grainstone”
ES caliza-“grainstone”
FR calcaire à texture “grainstone”
RE G
AB Grst

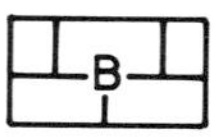

131. BOUNDSTONE: autochthonous carbonate in which original components are organically bound during deposition; boundstone is sometimes subdivided into framestone, bindstone and bafflestone.

NE boundstone
DE Boundstone
ES boundstone

ES boundstone
FR boundstone; roche construite
RE B
AB Bdst

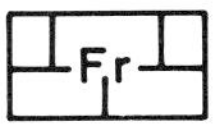

132. FRAMESTONE: a specific term, derived from Embry and Klovan's (1972) addition to Dunham's (1962) classification, denoting carbonates bound by organisms which formed a rigid framework (e.g., branching corals in which the skeletons of the organisms form a major element of the texture).

NE framestone
DE Framestone
ES framestone
FR framestone; roche construite

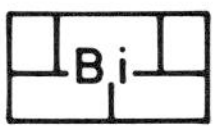

133. BINDSTONE: a specific term, derived from Embry and Klovan's (1972) addition to Dunham's (1962) classification, denoting a boundstone formed by organisms which encrust and bind the sediment but the skeletons of which need not be preserved (e.g., algal stromatolite).

NE bindstone
DE Bindstone
ES bindstone
FR bindstone; roche construite

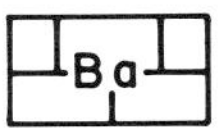

134. BAFFLESTONE: a specific term, derived from Embry and Klovan's (1972) addition to Dunham's (1962) classification, denoting a boundstone resulting from trapping of sediment by organisms which acted as baffles, locally reducing the rate of flow of water.

NE bafflestone
DE Bafflestone
ES bafflestone
FR bafflestone

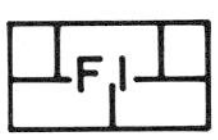

135. FLOATSTONE: a specific term, derived from Embry and Klovan's (1972) addition to Dunham's (1962) classification, denoting lime mud–supported carbonate rock with more than 10% particles larger than 2 mm in diameter (lime-wackestone is used when particles are smaller than 2 mm).

NE floatstone
DE Floatstone
ES floatstone
FR floatstone

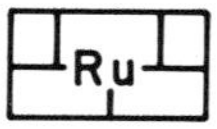

136. RUDSTONE: a specific term, derived from Embry and Klovan's (1972) addition to Dunham's (1962) classification, denoting grain-supported carbonate rock with particles larger than 2 mm in diameter (grainstone or packstone is used when particles are smaller than 2 mm).

NE rudstone
DE Rudstone
ES rudstone
FR rudstone

Definitions 137–153 (Examples of Monogranulate Carbonates)

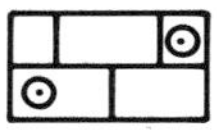

137. OOLITE: a sedimentary rock, frequently a carbonate, made up chiefly of ooids.

Note: See also Definition 58.

NE oöliet
DE Oolith; oolithischer Kalk
ES caliza oolítica
FR calcaire oolithique
RE ODRC
AB ool.

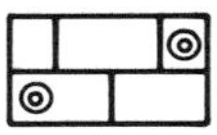

138. PISOLITE: a sedimentary rock, usually a carbonate, made up chiefly of pisoids.

Note: See also Definition 62.

NE pisoliet
DE Pisolith
ES caliza pisolítica
FR calcaire pisolithique
RE PDRC
Ab Pisol.

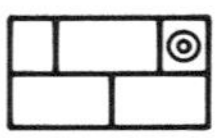

139. PISOLITIC: containing pisoids.

NE pisolitisch
DE pisolithisch
ES pisolítico; pisolítica
FR pisolitique; calcaire pisolithique
RE PDRC.
AB pisol.

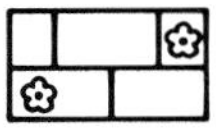
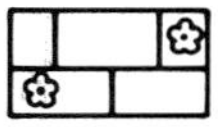

140. ONCOLITE: a sedimentary rock, usually a carbonate, made up chiefly of oncoids.

Note: See Definitions 64 and 163.

NE oncoliet
DE Onkolith
ES caliza oncolítica
FR calcaire oncolithique; calcaire à oncolites
RE OKRC
AB Onkl.

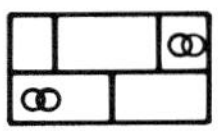

141. BAHAMITE (obsolete term): rock made predominantly of bahamiths (obsolete term). These consist of grains formed by smaller particles, predominantly peloids, which have been bound together by precipitated material to form aggregate grains. They grade in size from calcisiltite to calcirudite.

NE bahamiet
DE Bahamit
ES bahamita
FR bahamite
RE LURC
AB Gpst.

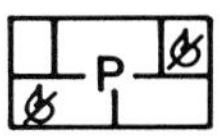

142. MICROCOQUINA: silt-sized detrital limestone partly made up of shells and/or shell fragments.

NE microcoquines
DE Mikroschillkalk;
ES microcoquina; microlumaquela
FR microlumachelle
RE SFRC
AB microcoq.

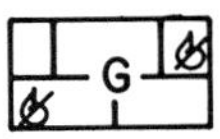
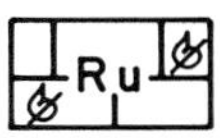

143. COQUINA: a limestone composed of loosely cemented shells and shell fragments.

NE schelpenkalk
DE Schillkalk
ES coquina; lumaquela
FR lumachelle
RE SFRC
AB Coq.

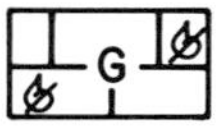

144. COQUINITE: a well-indurated and firmly cemented equivalent of coquina.

NE schelpenkalksteen
DE stark verfestigter Schillkalk
ES coquinita
FR lumachelle
RE SFRC
AB Coqt

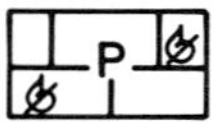

145. COQUINOID LIMESTONE (COQUINOID): a limestone consisting of coarse, often unbroken shelly material that has accumulated in place.

NE Schelpkalk
DE autochthoner Schillkalk;-Fossilkalk
ES caliza lumaquelica
FR calcaire lumachellique
RE SFRC
AB Coq. lst

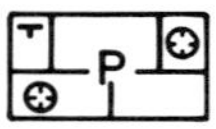

146. CORAL LIMESTONE: a limestone composed predominantly of corals (or coral fragments).

NE koraalkalk
DE Korallenkalk(stein)
ES caliza de corales
FR calcaire corallien
RE CRLS
AB Cor lst

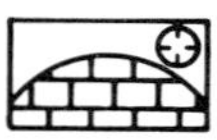

147. REEF LIMESTONE: a limestone consisting of the remains of reef-building organisms, such as corals, sponges and bryozoans, and of sediment-binding organic constituents, such as calcareous algae, in an association that is taken to be essentially in place (i.e., not a purely mechanical association of debris).

NE rifkalk
DE Riffkalk(stein)
ES caliza arrecifal
FR calcaire récifal
RE RLS
AB Rf lst

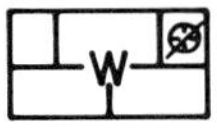

148. CORAL MUD: lime mud in which corals (or coral fragments) occur. The term is frequently used loosely.

NE koraalhoudende slib
DE korallenführender Schlamm
ES fango de corales
FR boue corallienne
RE CRCL
AB Cor md

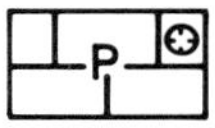

149. CORAL SAND: (loose) sediment consisting of sand-sized particles of corals.

NE koraalzand
DE Korallensand
ES arena de corales
FR sable corallien
RE CSSA
AB Cor S

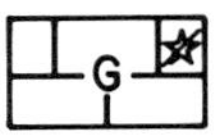

150. CRINOIDAL LIMESTONE: a limestone containing abundant crinoid fragments.

NE crinoiden kalksteen
DE Crinoidenkalk(stein)
ES caliza de crinoideos
FR calcaire à entroques; calcaire à crinoïdes
RE CSLS
AB Crin lst

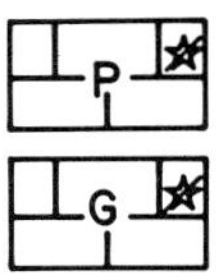

151. ENCRINAL LIMESTONE: a crinoidal limestone with more than 10% and less than 50% crinoidal fragments.

NE crinoidenarme kalksteen
DE crinoidenführender Kalk(stein)
ES caliza con crinoideos
FR calcaire à encrines
RE CSLS
AB (Crin) lst

152. ENCRINITIC LIMESTONE (ENCRINITE): crinoidal limestone with more than 50% crinoidal fragments.

NE crinoidenrijke kalksteen
DE Crinoidenkalk(stein); Trochitenkalk
ES caliza encrinítica
FR calcaire à encrines
RE CSLS
AB Crin lst

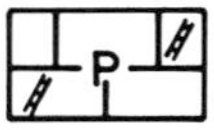

153. ALGAL LIMESTONE: a limestone formed by algal binding. Such binding is usually the result of the activity of blue-green algae.

NE algenkalk
DE Algenkalk(stein)
ES caliza de algas
FR calcaire à algues
RE AGLS
AB Alg lst

Definitions 154–156 (Examples of Hybrid Terms)

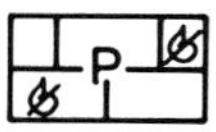

154. BIOCALCISILTITE: a calcisiltite containing abundant bioclasts.

NE kalkslibsteen met bioklasten
DE Biokalksiltit
ES biocalcisiltita
FR biocalcisiltite
RE SJSF
AB slt, biocalc

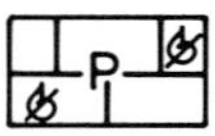

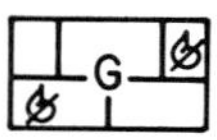

155. BIOCALCARENITE: a calcarenite containing abundant bioclasts.

NE kalkzandsteen van skeletfragmenten
DE Biokalkarenit
ES biocalcarenita
FR biocalcarénite
RE ARSF
AB sst, biocalc.

156. BIOCALCIRUDITE: a calcirudite containing abundant bioclasts.

NE kalkkiezelsteen met skeletfragmenten
DE Biokalkrudit
ES biocalcirudita
FR biocalcirudite
RE RUSF
AB Rud, biocalc.

1–5 CARBONATE DEPOSITIONAL STRUCTURES

Introduction

Depositional sedimentary structures are features of a rock which are the result of physical, chemical and biological processes, active in the depositional environment, such as sorting and/or orientation of sedimentary grains with respect to size, shape and specific gravity. These processes leave products that provide information on the medium of deposition and on the energy conditions present in it. Depositional structures are such products. In carbonates they have been subdivided into mechanical, organomechanical and organic structures, based on their formative agents. They are usually better studied in core or outcrop than in thin section. Many of the structures are identical to those found in siliciclastic environments, where similar environmental processes may act.

Depositional Structures Formed by Mechanical Processes

Inorganic primary structures are organised into surface and bedding features (Fig. 1–15). Linear features are considered external bedding features.

Bedding Features. Sedimentary deposits are characterized by stratification. Units of stratification are beds and laminae. Their distinction is based on thickness (Table 1–2). They are not mutually exclusive. The term *massive* best describes a thick bed void of internal structures. Beds which have internally a progressive decrease or increase in grain size upwards are termed *graded.* Fining upwards of grain size is characteristic of mechanical deposition under increasingly quieter water conditions, whereas coarsening upwards indicates a shift towards high-energy depositional conditions.

Very thin (graded) laminations are common in shallow marine carbonates and evaporites, but these are not always exclusively mechanical in origin (see, e.g., Hardie, 1977) (see also p. 60).

Inclined Stratification. Inclined strata are the result of sand-wave migration or of accretion on an inclined surface. Descriptions of types and discussions on the origin of inclined strata can be found in scores of modern textbooks, mainly those on siliciclastic deposits (e.g., Reineck and Singh, 1973). Most of the types in Fig. 1–15 also occur in Recent or ancient carbonates.

DEPOSITIONAL	MECHANICAL	BEDDING FEATURES	STRUCTURELESS			
			HORIZONTAL BED	see p. 57		
			HORIZONTAL LAMINAE			
			GRADED BED			
			Inclined Bed	Migratory Sand Waves	Festoon	Forest: Small Scale < 10 cm
						Meduim scale 10 cm – 1 m
						Large scale > 1 m
						Backset
					Lobate	
				Accretion to Stationary Slope	Lateral Accretion	Accretional beach foreslope
						Accretional bank slope
						Accretional channel fill
						Accretional bar
					Spillover	Spillover channel fill
						Bar spillover
						Spillover delta
						Spillover tidal bar
						Spillover dune
						Beach backslope washover
		TOP SURFACE	RIPPLE MARKS	Parallel	Symmetrical	
					Asymmetrical	
				Scalloped	Crescentic	
					Lobate	
		BOTTOM MARKINGS		Flute casts		
		LINEATIONS		Lineated sand grain		
				Lineated plant fragments		
				Lineated shells		
				Lineated pebbles		
		Structures in reefs		Framestone, bindstone, bafflestone		
		Stromatolitic structures		Stromatolite		
				Thrombolite		

Fig. 1–15. Common depositional sedimentary structures in carbonates.

Table 1–2
Classification of Bedding Thickness

Shell's standard legend	REGEO (computer) code	Classification of Reineck and Singh (1973)
mm-Bedded, <1 cm	BB, (mm)	Very thin
cm-Bedded, 1–10 cm	BC, (cm)	Thin
dm-Bedded, 10–(30) cm	BD, (dm)	Medium
(30)–100 cm		Thick
m-Bedded, >100 cm	BM, (m)	Very thick

External Bedding Features. Erosional linear structures are usually bottom markings. Lineations in carbonates are formed by shell (fragment) alignment. Neither erosional nor lineation structures are common in Recent or ancient carbonates.

Depositional Sedimentary Structures Formed by Organomechanical or Organic Processes

Cemented accumulations formed as a result of local organic productivity or selective sediment trapping by organisms can have mounded, undulating or irregular bedding surfaces that are very different from flat and parallel surfaces produced by water movements. Such undulations range from a fraction of a centimetre to many kilometres in extent. Thus there is need for an arbitrary subdivision into depositional sedimentary structures, as discussed here and in relation to sedimentary bodies (Section 2–4). A practical means of separating the two is the dimension of a single outcrop or a core. Distinction should also be made between sedimentary structures or bodies and grains. A structure or body maintains a fixed position with respect to the substrate, which is not the case with grains. Consequently oncoids are grains and not structures. Some depositional structures formed by organomechanical or by mechanical processes are listed in Fig. 1–15.

Structures in Ecological Reefs. Frame-building, sediment-forming and sediment-binding activities of organisms form together an autochthonous carbonate mass—the ecological reef. Reefs proper are depositional bodies and they are not structures (see above and Chapter 2). The various actions of organisms in the reef could give rise to such structures as baffle, bind and frame structures. The resulting depositional textures are bafflestones, bindstones and framestones (Embry and Klovan, 1972). These depositional–textural terms (see also p. 47–48) have thus a genetic significance.

Stromatolitic Structures. Carbonate masses composed of calcareous laminae suggesting the former presence of sediment-binding algae are stromatolitic structures (Logan *et al.*, 1964). Two basic geometric forms occur: laterally linked hemispheroids (LLH) and stacked hemispheroids (SH). A third form, spheroidal (SS), is disconnected from the sediment mass and is consequently discussed as a grain.

Form diversity in stromatolites is thought to reflect physical environmental factors, notably degree of insolation. Algae grow towards the sun, which gives rise to undulating stromatolites and, in extreme cases, to algal pillars. The term *algal limestone* is often used to indicate limestone bodies with stromatolitic structures. This is incorrect; such limestone should be termed stromatolitic. It should be kept in mind, however, that there is no direct evidence of a relationship between fine laminations and stromatolitic structures unless additional features such as those mentioned above are present. Together with mud cracks and flat pebble conglomerates, stromatolites may characterise inter- to supratidal settings. Once broken down they often give rise to rubble beds. Thrombolites are problematic cryptalgal structures.

Origin and Significance of Depositional Sedimentary Structures

Identical depositional structures may be formed by comparable or by quite different processes in various depositional environments. Lamination, for instance, could reflect mechanical, organomechanical or purely organic processes such as, respectively, settling of pelagic sediment, stromatolite formation and travertine precipitation. Thus it may occur in deep marine, intertidal or terrestrial environments.

Laminations are not equally well preserved in each of the different environments. Good preservation requires an absence of burrowers and bottom feeders. These conditions can be met (1) in areas with frequent and rapid deposition in euxinic environments, (2) in hypersaline (evaporitive) environments and (3) in inter- or supratidal depositional areas. Thus sedimentary laminations are usually preserved within the following environments:

1. Deep marine
   a. Areas with stagnant bottom waters such as depressions, troughs and trenches on the sea floor
   b. Areas of frequent, rapid deposition
2. Shallow marine
   a. Hypersaline lagoons, lakes, pools, bays, estuaries
   b. Local depressions (catchment areas for fine sediment)
3. Supratidal
   Tidal flats

Distinction of such laminations can be achieved only by taking into account the associated sedimentary features that could provide criteria for the recognition of the depositional environment. Directional structures are scarce in carbonates. When present, they may indicate regional trends in carbonate bodies. As carbonate grains of equal size do not necessarily have the same hydrodynamic significance, an interpretation of mechanical sedimentary structures in carbonates can often be a hazardous undertaking. But, as with laminae, an interpretation of the association of sedimentary structures may lead towards an interpretation of the particular combination of processes which must have been operative during deposition.

Definitions 157–174

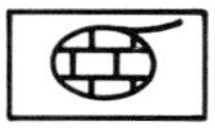

157. RESEDIMENTED CLASTS; MÉLANGE: beds containing slumped or otherwise dislocated masses of carbonate rocks. Original stratification may still be visible, or the whole mass may consist of a mixture (mélange) of boulders or larger blocks with no internal deformation.

NE opnieuw afgezette brokstukken; mengsel
DE allochthone Karbonate
ES clastos resedimentados; mélange

FR éléments exotiques resédimentés
RE OHLS
AB Lst Olisth

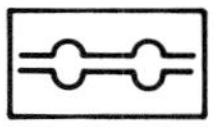

158. BALL AND FLOW; SEDIMENTARY BOUDINAGE; BALL AND PILLOW: differential compaction of originally patchily distributed (flaser-bedded) shale and carbonate, resulting in irregular, closely spaced, almost nodular structures. Disruption of layers caused by stretching and flowage.

NE vloeistructuur; sedimentaire boudinage
DE Boudinage Struktur
ES "Ball and Flow"; boudinage sedimentario
FR miche; boudinage; bouler et reiller
RE BN

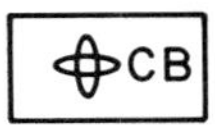

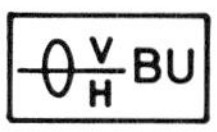

159. BIOTURBATION; BURROWS: reworking of sediment by organisms. Individual organism traces preserved through colour mottling and textural differences, or outlined by dolomitisation.

Note: For classification of these structures see Seilacher (1964).

NE organische doorwoeling; graafgangen
DE Bioturbation; Verwühlung
ES bioturbacíon
FR bioturbation; terriers
RE BU, CB
AB Bur

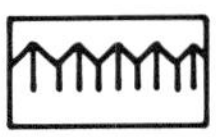

160. CORRASION PLANE: abrasion and truncation of surfaces by streaming water, by dissolution by meteoric water or by organic activity. The structure is often indistinguishable from "hardgrounds".

NE erosievlak
DE Abrasionsflächen
ES superficie de corrasion
FR surface d'abrasion; surface de dissolution; surface de karstification

161. BLACKENED PEBBLES: blackened particles due to algal activity. They might indicate subaerial exposure in areas filled periodically with hypersaline water. They also may indicate the presence of hydrotroillite (a precursor of pyrite), precipitated in a reducing environment in which decaying organic matter occurs. They may be reworked into overlying beds.

NE zwarte kiezels
DE "Black Pebbles"
ES cantos ennegrecidos
FR cailloux noirs
RE PBBK
AB Pbl, blk.

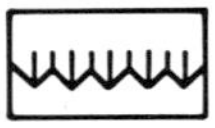

162. STORM LAYER: widespread, commonly unfossiliferous and well-marked layers resulting from major storms, deposited when rising tide and high winds mobilise mud on bottoms of normally quiescent shallow ponds and lakes and deposit thick layers on shoal areas.

Note: Worthwhile reviews on various aspects of storm layers are given by Einsele and Seilacher (1982).

NE stormlaag
DE Sturmflutsediment; Tempestit
ES capa con depósito de tormenta
FR depôt de tempête; cordon de tempête

163. ALGAL STRUCTURE: a sedimentary structure of definite form resulting from carbonate secretion and precipitation by algal colonies. This includes crusts, small (pseudo)pisolitic and pseudoconcretionary forms, biscuit- and cabbagelike heads of considerable size, laminated structures such as stromatolites or bedding modified by blue-green algal mats. Some so-called algal structures may be of inorganic origin.

Note
1. See Monty (1981).
2. In the German and French languages the term *texture* and *structure* are used inversely from the English.
3. See also Definitions 64 and 140.

NE algenstructuur
DE Algenstructuren
ES estructura de algas
FR structure algaire
RE SM

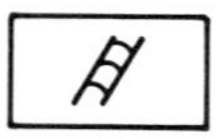

164. CALCAREOUS ALGAE: a taxonomically heterogeneous group of predominantly marine algae that secrete calcium carbonate.

NE kalkalgen
DE Kalkalgen

ES algas calcáreas
FR algues calcaires
RE AG
AB Alg

165. PROTOSTROMATOLITE: unlithified laminated structure produced by blue-green algae.

NE beginnend stromatoliet
DE Protostromatolith
ES protoestromatolito
FR proto-stromatolite
RE SMPR

166. STROMATOLITE: a variously shaped, laminated, mainly calcareous sedimentary structure, attached to the substrate and formed by blue-green algae or bacteria, which form an algal or bacterial mat that bind fine detritus and/or precipitated calcium carbonate biochemically (see Fig. 1–16).

Note: See also Walter (1976) and Monty (1981).

NE stromatoliet
DE Stromatolith
ES estromatolito
FR stromatolite
RE SM

167. SPONGIOSTROME STROMATOLITE: calcified, tiny bush-tufted algal mats showing up as anastomosing filamentous structures.

NE sponsachtige stromatoliet
DE Spongiostromate Algen-Matten
ES spongiostrome stromatolite
FR stromatolite à filaments ramifiés

168. BULBOUS STROMATOLITE: laterally linked hemispheroidal stromatolites, commonly formed by Collenia or cryptozoan algae. The bulbous form results from continuous and/or intermittent algal growth, possibly over original protrusions of substrate such as arches or mud-cracked polygons (see Fig. 1–17).

Note: See Logan *et al.* (1964).

NE knolvormige stromatoliet
DE knolliger Stromatolith

STRUCTURAL FORMULAE	DESCRIPTION	VERTICAL SECTION OF STROMATOLITE STRUCTURES
$\frac{\text{LLH-S}}{\text{LLH-C}}$	Space-linked hemispheroids with close-linked hemispheroids as a microstructure in the constituent laminae.	
$\frac{\text{SH-V}}{\text{LLH-C}}$	Discrete, vertically stacked hemispheroids composed of close-linked hemispheroidal laminae on a microscale.	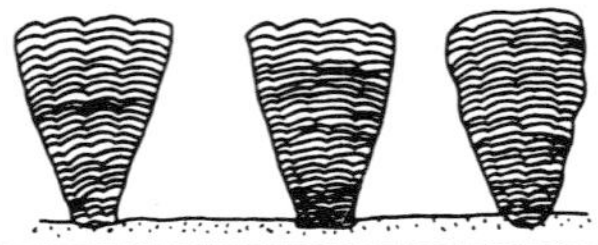
LLH-S→SH-V	Initial space-linked hemispheroids passing into discrete, vertically stacked hemispheroids with upward growth of structures and continued differentiation of domes and interareas.	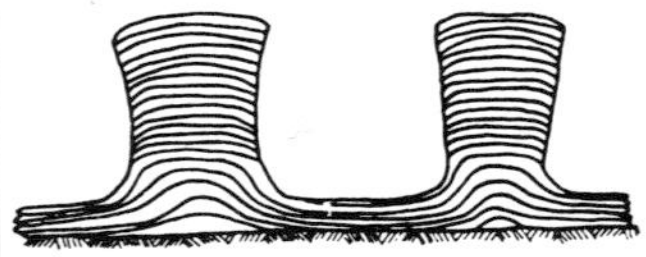
SH-V→LLH-C	Initial discrete, vertically stacked hemispheroids passing into close-linked hemispheroids by upward growth expansion.	
SH-V→LLH-S→SH-V→LLH-S→SH-V	Alternation of discrete, vertically stacked hemispheroids and spaced-linked hemispheroids due to periodic sediment infilling of interstructure spaces.	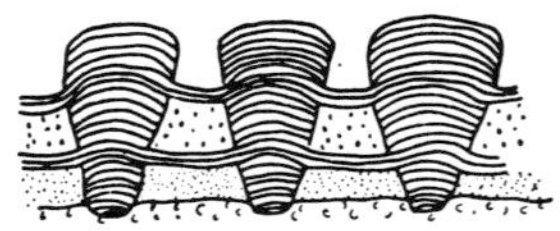
$\frac{\text{LLH-S}\rightarrow\text{SH-V}}{\text{LLH-C}}$	Inital space-linked hemispheroids passing into discrete, vertically stacked hemispheroids; both with laminae of close-linked hemispheroids.	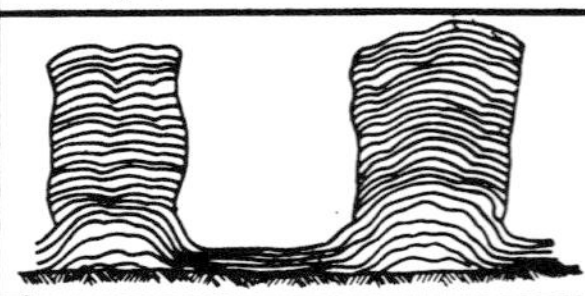
$\frac{\text{SH-V}\rightarrow\text{LLH-C}}{\text{LLH-C}}$	Initial discrete, vertically stacked hemispheroids passing into close-linked hemispheroids; both with laminae of close-linked hemispheroids.	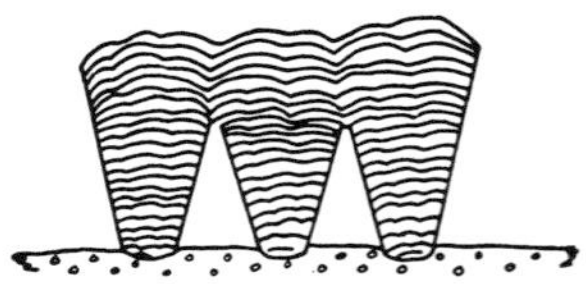
$\frac{\text{SS-C}}{\text{LLH-C}}$	Concentrically stacked spheroids with laminae composed of close-linked hemispheroids.	

Fig. 1–16. Compound stromatolite structures. Illustration by permission of The University of Chicago Press.

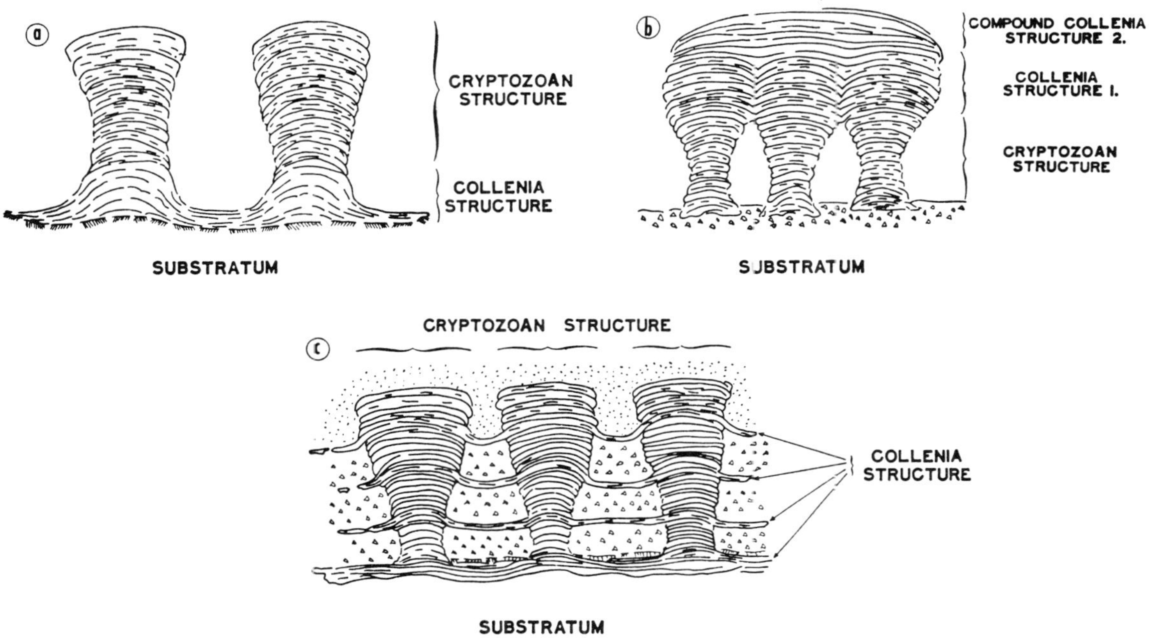

Fig. 1–17. Vertical cross section of compound stromatolitic structures. Illustration by permission of the University of Chicago Press.

ES estromatolita bulbosa
FR stromatolite en hémisphères
RE SMDO

169. THROMBOLITE: non-laminated, cryptalgal stromatolite, probably formed in shallow marine to low intertidal depositional environments, but also reported in deep-water settings.

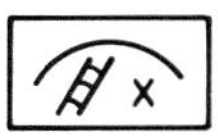

Note: See Aitken (1967) and Rubin and Friedman (1977). Thrombolite–stromatolite buildings have also been described from the Mesozoic in deep, quiet water settings (Ellis *et al.*, 1985).

Ne thromboliet
DE Thrombolit; Krypto-Algen-Gefüge
ES trombolito
FR thrombolite

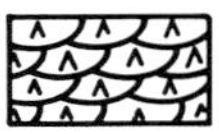

170. CHICKEN-WIRE STRUCTURE: nodular structure formed generally by compaction of calcium sulfate nodules in sediment of varying permeability. Such structures are seen today forming on modern sabkhas but also result from compaction following burial.

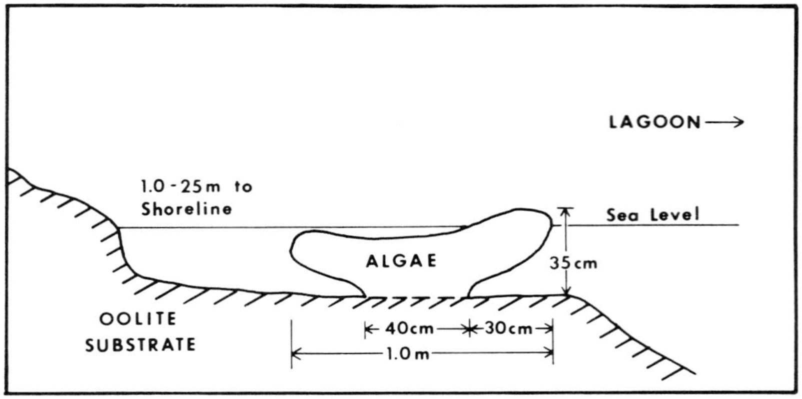

Fig. 1–18. Generalized cross section of an individual algal head with typical measured dimensions. Illustration by permission of the Soc. Econ. Palaeontol. Mineral.

NE kippegaas structuur
DE Knollengefüge
ES estructura "chickenwire"
FR structure "cage à poule"; structure grillagée

171. ALGAL HEAD: a bulbous, dome-shaped or columnar mass of mechanically transported, laminated sediments collected by algae (especially blue-green algae) on a tidal flat or in a lake and bound together by algal filaments (see Fig. 1–18).

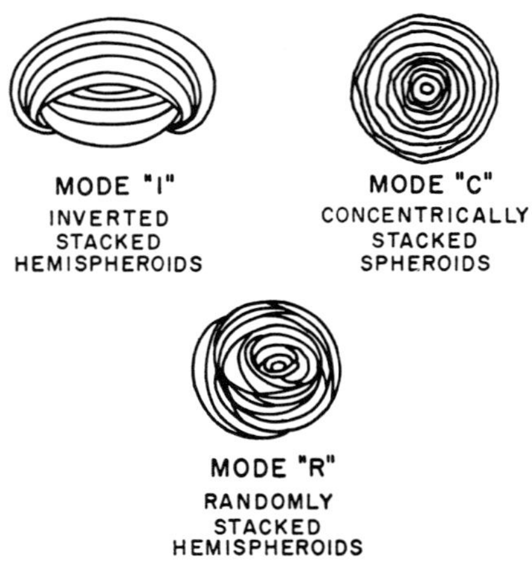

Fig. 1–19. Spheroidal stromatolitic structures (type SS). Illustration by permission of the University of Chicago Press.

Note: See Thornton *et al.* (1978)

NE algendome; algenknobbel
DE Algenkuppel
ES —
FR édifice stromatolitique; édifice algaire

172. ALGAL BALL; ALGAL BISCUIT: hemispherical or disk-shaped calcareous masses up to 20 cm in diameter and often concentrically laminated (see Fig. 1–19).

Note: Algal balls or biscuits can be both skeletal (then termed rhodolith) and stromatolitic.

NE algenbal
DE Algenknolle
ES "algal ball"
FR boulet algaire; oncolithe; biscuit d'eau

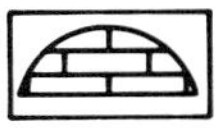

173. ALGAL MOUND: heads or knolls made of algal limestone.

NE algenheuveltje
DE Algenhügel
ES "algal mound"
FR dôme algaire

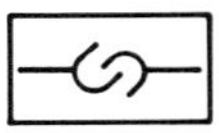

174. SOFT SEDIMENT SLUMPS: contorted or folded bedding which is restricted to an individual bed and for which there is no evident tectonic cause.

NE sedimentaire afglijdingen
DE Schlammgleitung
ES —
FR lits deformés; lits plissés; lits slumpés
RE BS
AB slump

2

Carbonate Environments, Settings and Bodies

2–1 INTRODUCTION

The environmental factors that govern the formation and distribution of carbonates differ greatly from those controlling siliciclastics (Fig. 2–1). This makes it difficult to compare these factors. Consequently the environmental arrangement suggested in this chapter is applicable to carbonates but not necessarily in the same way to siliciclastics. Examples of significant differences between the two realms of sedimentation are outlined below. Approximately half of the carbonates preserved in the geological record have been deposited in shallow water. Thus a subdivision of environments defined exclusively in terms of water depth is less meaningful for carbonates than it might be for siliciclastics. Such factors as oxygenation, salinity, temperature and water circulation co-determine the abundance and nature of pelagic and benthonic life, and skeletons of organisms with these life habitats form the bulk of carbonate sediments. Thus these factors could be used to make a meaningful subdivision of carbonates deposited within the shallow marine zone. Such factors are only partly depth related; they are also influenced by climate, the configuration of the seashore and that of the sea bottom. Many kinds of organic calcareous deposits, in particular those formed by skeletons of benthonic organisms, are approximately formed in place. However, deposits of mechanically transported carbonate particles are being more and more commonly recognised [Cook *et al.* (1983); Crevello and Harris (1985); Cook and Mullins (1983)]. Turbidity of water due to suspension of carbonate mud or clay minerals suppresses or even prevents the production of carbonates, and hence carbonate accumulation, because it inhibits photosynthesis, whereas it certainly does not prevent coarser siliciclastics from accumulating.

With this in mind an outline has been made (Fig. 2–2) in which a reconciliation of these factors is attempted. Moreover, a hierarchical arrangement—from bigger entities on the left to smaller ones on the right—has been maintained. The main carbonate environmental settings are terrestrial, coastal–marine and marine. The transitional coastal–marine environmental setting is further divided into supratidal and intertidal zones. In the figure, sediment bodies occurring in both the terrestrial and the coastal–marine settings have been distinguished according to their preferred

DEEP MARINE		SHALLOW MARINE	INTERTIDAL	SUPRATIDAL
DEEPEST MARINE	DEEPER MARINE			
BELOW THERMOCLINE; CARBONATE DISSOLUTION WITH DEPTH	BELOW EFFECTIVE WAVE BASE OR BASE OF EUPHOTIC ZONE; NO CARBONATE DISSOLUTION EFFECTS WITH DEPTH	BETWEEN LOW TIDE AND EFFECTIVE WAVE BASE OR BASE OF EUPHOTIC ZONE	BETWEEN HIGH TIDE AND LOW TIDE	BETWEEN HIGH TIDE AND STORM TIDE

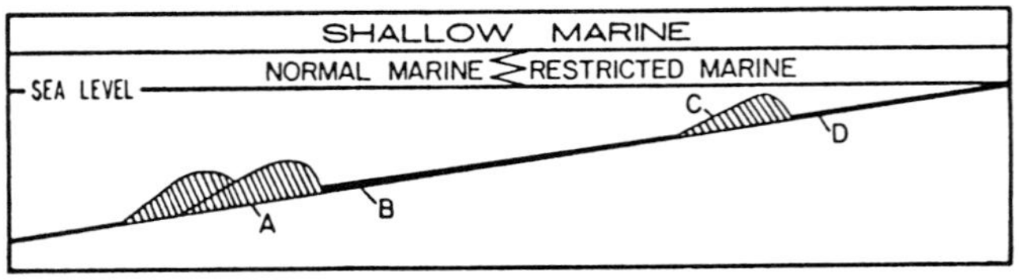

Fig. 2–1. (Top) Bathymetric subdivision of carbonates. (Bottom) Environments on a very wide, low-gradient slope. A, Normal shallow marine, high energy. B, Normal shallow marine, low energy. C, Restricted shallow marine, high energy (caused by wind-generated currents). D, Restricted shallow marine, low energy.

occurrence, either in humid tropical–subtropical or in arid (sub)tropical climate zones. The outline of Fig. 2–2 serves as a guide to the sequence of defined terms compiled in this chapter. It is realised that an overall hierarchical classification of such a variety of environments, settings and carbonate bodies is open to addition and refinement. It is felt, however, that Fig. 2–2 and the subsequent thematic arrangement of terms and their definitions give a fair overview of the field of carbonate environments and sediment bodies.

It the present chapter definitions of general terms related to carbonate depositional environments and facies precede a systematic discussion of individual depositional settings and bodies. The terms related to marine bodies and settings can be grouped into those related to reefs and those related to lagoons. Further subdivisions have been made within each section. The terms related to tidal–terrestrial sedimentary bodies and settings have been treated in one group. A discussion of ancient carbonate depositional settings and bodies and a comparison with those in the Recent conclude the chapter.

2–2 CARBONATE DEPOSITIONAL ENVIRONMENTS

Introduction

The awareness that environments* are not quite the same as rocks, which are tangible, nor the settings through which one can walk, but rather comprise both the environmental conditions and

* The term *environment* is repeatedly found in the literature in the sense of "settings", "body" or "surrounding". In the present series of definitions it is used exclusively in the sense of "surrounding" (the French *milieu*).

<table>
<tr><td>Main environment</td><td colspan="4">Specific carbonate setting</td></tr>
<tr><td>With respect to wave base or light penetration</td><td>With respect to tidal range</td><td>With respect to climate</td><td>With respect to geomorphic forms</td><td>With respect to discreet carbonate bodies</td></tr>
<tr><td rowspan="2">Terrestrial, T</td><td rowspan="4">Supratidal, CT</td><td>Humid tropical</td><td>Lake, TA</td><td rowspan="2">Palustrine limestones

Dune, D</td></tr>
<tr><td>Arid tropical</td><td>Calcrete plain, TC
Inland sabkha, TS
Saline desert, TD
Lake, TA
Bioclastic-ooid dune, D</td></tr>
<tr><td rowspan="4">Transitional marine to coastal "tidal", C</td><td>Humid tropical</td><td>Normal salt marsh, CK</td><td rowspan="3">Washover fan, E
Tidal bar, S
Spit, S
Beach ridge, B</td></tr>
<tr><td>Arid tropical</td><td>Coastal sabkha, CS
Evaporitic flat, CE
Supratidal flat, CT</td></tr>
<tr><td rowspan="2">Intertidal, CA</td><td>Humid tropical</td><td>Coastal pond, CA and lagoon, CL
Tidal flat, CE
Tidal channel, CC
Tidal delta, CF
Beach, CB</td></tr>
<tr><td>Arid tropical</td><td>Evaporitic coastal pond, CA and lagoon, CL
Tidal flat, CE
Tidal channel, CC</td><td>Oolitic tidal bar, S</td></tr>
<tr><td rowspan="2">Full marine, M</td><td>Shallow marine, MS</td><td></td><td>Reef flat, MR
Reef wall, MW
Offshore bank, MS
Outer shelf-bank margin, MM</td><td>Bioclastic - ooid sand bar - shoal, SC
Shelf reef, RS
Patch reef, RP
Fringing - barrier reef, RF/RB
Shell bank, SS
Winnowed - lag, G
Sedimentary blanket, EB
Lime mud mound, PM</td></tr>
<tr><td>Deep(er) marine, MD</td><td colspan="2">With respect to O_2 content and circulation
- - - - - - - - - - - - - - - - - -
Normal salinity basin plain and O_2, OB
Euxinic, E
Ocean bottom, OB
Evaporitic, EV</td><td>Reef talus, RE
Gravity slide mass, XG
Slump mass, X
Olistostrome, Y
Turbidite, T
Pelagic ooze sheet, Z</td></tr>
</table>

Fig. 2–2. Carbonate sedimentary environments and settings.

the setting, has guided the arrangement and definition of the terms that follow. Critical parameters in subdividing the environments are depth, hydraulic energy level and salinity. Environmental analysis starts with observations based on which interpretation can be attempted, leading to a better understanding of the environment. A restricted number of sedimentary parameters can selectively be looked for (Fig. 2–3), and such a restrained approach does allow the recognition of major groups of depositional environments. In addition such recognition often allows interpretations pertaining to the setting in which the sediment bodies are formed. This helps one to establish a conceptual depositional model with the aid of which one can extrapolate the possible areal extent of environmental zones in which characteristic sediments come to rest.

Definitions 175–184

175. (CARBONATE) DEPOSITIONAL ENVIRONMENT: the combination of all the surrounding conditions, influences or forces affecting a locus of (carbonate) sedimentation.

Note: Environmental subdivision is carried out by considering the combined influences on the carbonate sediments of the geochemistry of the water, the tides, the effective wave base, the presence or absence in the photic zone and the presence or absence in the depth zones in which carbonate minerals dissolve. Based on this a subdivision can be made into supratidal, intertidal, shallow marine and deep marine environments.

NE kalkafzettings milieu
DE Ablagerungsbereich; Karbonatablagerungsmilieu
ES ambiente carbonatado; medio carbonatado
FR environnement carbonaté
RE EN

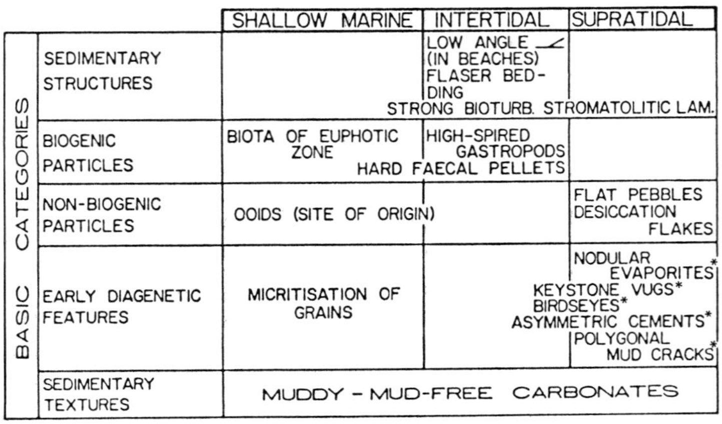

Fig. 2–3. Diagnostic features of shallow-water carbonates. For an explanation and further details on the carbonates marked with an asterisk, see Chapter 3.

176. EXPOSURE INDEX (EI): at any one point on the sediment surface within the tidal range, the mean annual percentage of time the point is exposed (see Fig. 2–4).

Note

1. The EI is a relative but quantitative measure that allows precise comparison of the tide factors from one area to another, regardless of topographic or hydraulic differences (see Ginsburg *et al.*, 1970).
2. In the examples given in Fig. 2–4 (taken from Hardie, 1977) the environmental concept is different. These examples refer to "subenvironments", which in the present book are described as settings (e.g., levee backslope, channel floor).

NE droogvallingsindex
DE Trochenfalle-Index
ES indice de exposión
FR coefficient d'émersion

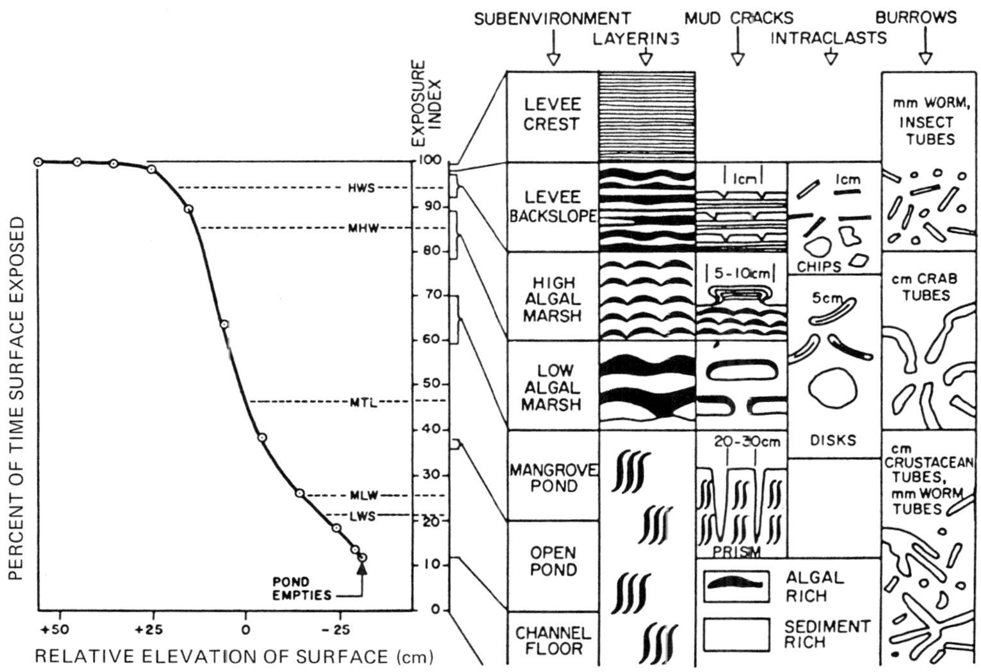

A

Fig. 2–4. A, Vertical distribution of sedimentary features of levee–pond subenvironments based on work by Hardy (1977). B, Common organisms of pond and level, plotted against exposure index. Based on work by Hardie (1977).

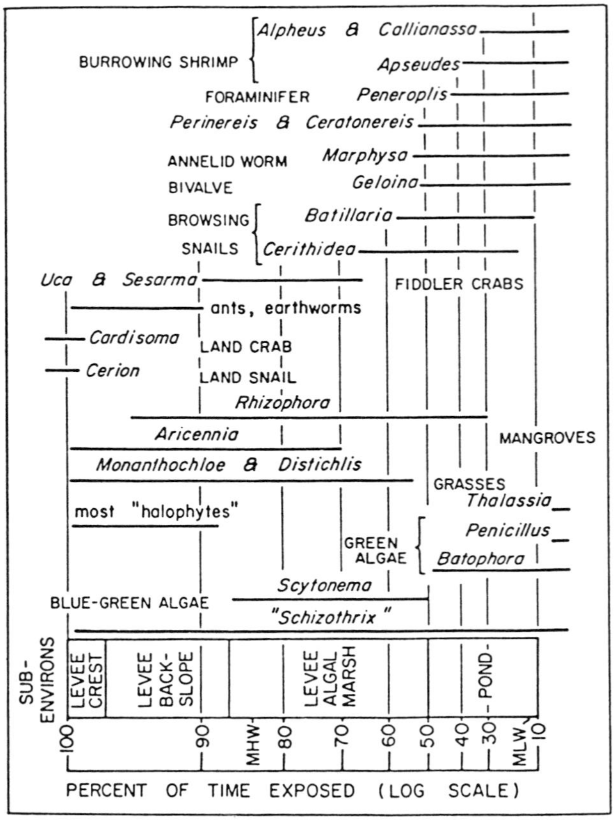

B

Fig. 2–4. (*Continued*)

177. (CARBONATE) SUPRATIDAL (ENVIRONMENT): encompasses the zone between high tide and storm tide.

NE karbonaat stormtij milieu
DE Supratidal; Spritzwasser-Zone
ES supramareal
FR supratidal
RE CT

178. (CARBONATE) INTERTIDAL (ENVIRONMENT): encompasses the zone between high tide and low tide.

NE karbonaat getijdenmilieu
DE Intertidal; Gezeiten-Zone; Watt-Zone

ES intermareal
FR intertidal
RE CA

179. (CARBONATE) DEEP MARINE (ENVIRONMENT): encompasses the entire zone below the wave base or the photic zone. On the basis of the presence or absence of carbonate dissolution effects (with depth), it might be possible to subdivide this environment into a deeper and a deepest zone (see Figs. 2–5 and 2–6).

NE diep-marien karbonaat milieu
DE Tiefwasser-Bereiche
ES marino profundo
FR milieu marin profond carbonaté
RE MD

180. ARAGONITE COMPENSATION DEPTH (ACD): that depth in the deep marine carbonate depositional environment at which aragonite dissolves.

Note: Calcite compensation depth (CCD) is similarly defined.

NE aragoniet compensatie diepte
DE Aragonitkompensationstiefe
ES profundidad de compensación del aragonito
FR profondeur de compensation de l'aragonite

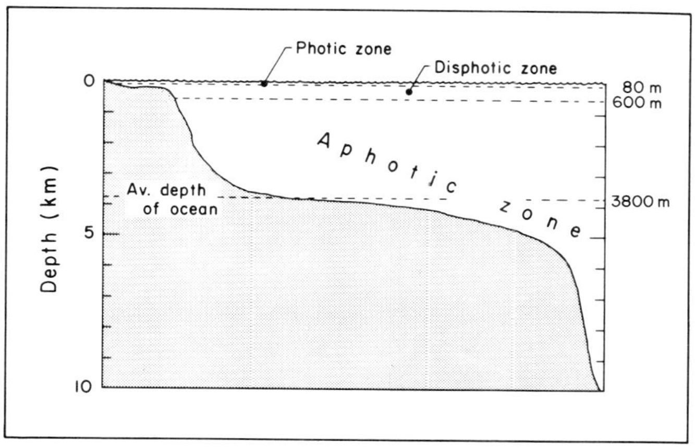

Fig. 2–5. Relationships of three major depth zones in the oceans, based on light penetration compared with schematic profiled ocean depth. Illustration after Hedgpeth (1957) by permission of the Geol. Soc. Am.

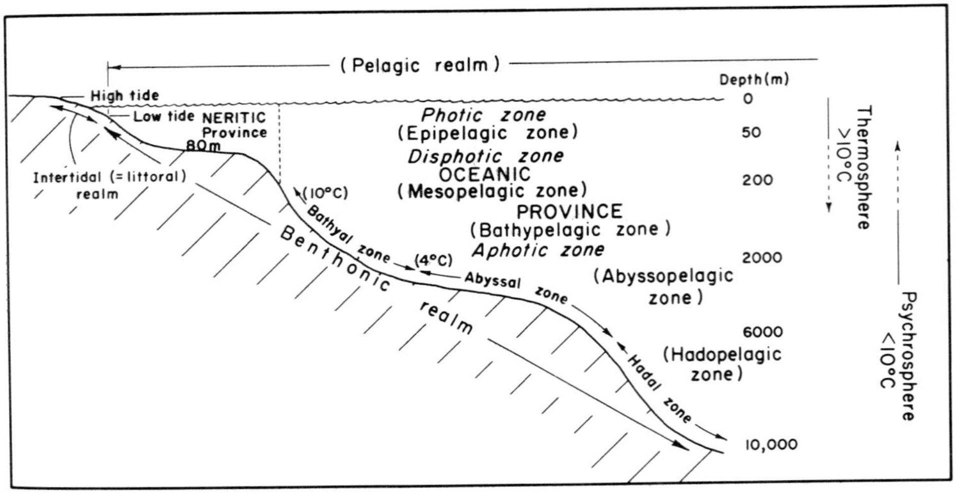

Fig. 2–6. Schematic profile through ocean showing ecological subdivisions of bottom and of water. Important boundaries are based on depth of penetration of sunlight, depth where water temperature becomes 10°C and depth below which water temperature is always 4°C or lower. Illustration slightly modified after Hedgpath (1957) by permission of the Geol. Soc. Am.

181. HIGH-ENERGY (CARBONATE) ENVIRONMENT: the shallow marine and intertidal carbonate environments which are not protected against wave and current agitation (see Fig. 2–7).

Note: Currents may also affect carbonate sediments in the deep marine environment, in which case such environments are also high-energy ones. They are rare, however, in comparison with the shallow marine high-energy environments.

NE hoog-energetisch karbonaat milieu
DE Hochenergetischer Ablagerungsbereich
ES medio de alta energía
FR environnement à haute énergie
RE EENH

182. LOW-ENERGY (CARBONATE) ENVIRONMENT: the shallow marine and intertidal carbonate environments which are protected against wave and current agitation.

Note: Deep marine environments are also usually low-energy environments. In rare cases, however, currents change such environments locally to high-energy ones.

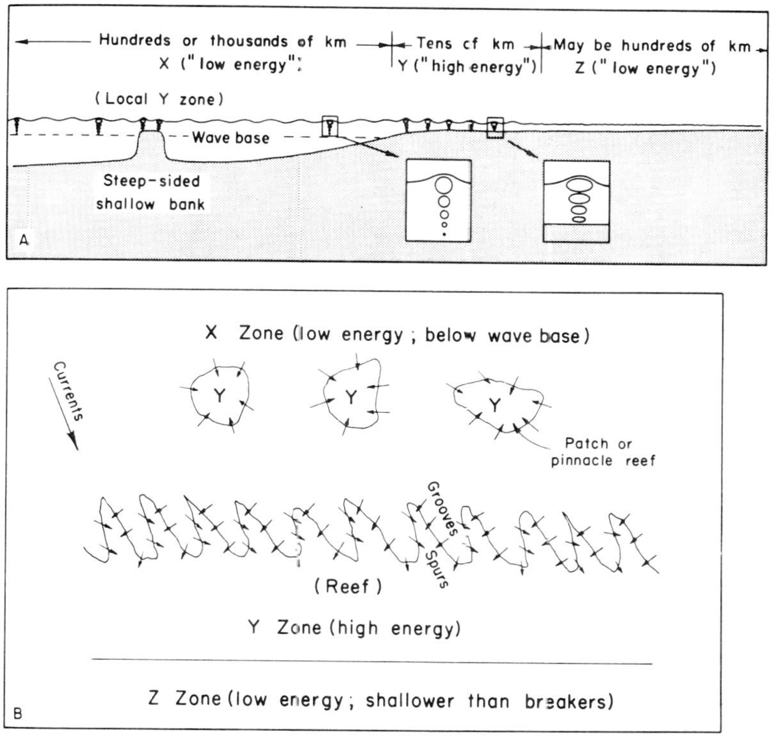

Fig. 2–7. Three schematic energy belts: X, Y and Z. A, Schematic profile showing depth relationships of energy belts to waves. In X zone is the bottom below the wave base; deep-water waves cause water "particles" to move in circular orbits. In Y zone is the bottom above the wave base, and water "particles" move in elliptical orbits. B, Sketch map of X, Y and Z zones. Reefs occur in Y zone; three pinnacle reefs occur on windward side of barrier reef. Illustration modified after Irwin (1965) by permission of the AAPG.

NE laag-energetisch karbonaat milieu
DE Niedrichenergetisher Ablagerungsbereich
ES medio de baja energía
FR environnement à basse énergie
RE EENL

183. NORMAL MARINE CARBONATE ENVIRONMENT: a carbonate environment characterised by water with a normal open oceanic salinity range (33–40‰) as evidenced by its effects on organisms.

NE marien karbonaatmilieu met een normale flora/fauna diversiteit
DE Normalmarines Milieu
ES ambiente carbonatado marino normal

FR milieu marin carbonaté normal
FR milieu marin carbonaté à salinité normale
RE RENO

184. RESTRICTED MARINE CARBONATE ENVIRONMENT: a carbonate environment characterised by water with a salinity range markedly different from that of open oceanic water (33–40‰), as evidenced by its effects on organisms and the precipitation of certain minerals (e.g., gypsum).

Note: The term *restriction* was originally used for "restricted" biological diversity, but it is also used to indicate "restricted" water circulation and consequently—under the right climatological conditions—evaporation, lack of oxygen, phosphorus, NO_3, etc.

NE marien karbonaat milieu met beperkte flora/fauna diversiteit
DE Eingeschränktes Milieu; restriktives Milieu
ES ambiente carbonatado marino restringido
FR milieu marin carbonaté à salinité variable
RE RENR

2–3 CARBONATE DEPOSITIONAL FACIES

Introduction

Facies in carbonate rocks are characteristic assemblages of sedimentological, palaeontological, petrographical and chemical features, reflecting processes active in the depositional and diagenetic environments. Thus the following groups of facies are recognisable:

1. Depositional
   a. Lithofacies
   b. Biofacies
2. Diagenetic
   a. Environment related
   b. Shallow burial
   c. Deep burial

Facies are studied at various levels of detail. The stereoscan and cathode-lumniscence microscope allow the investigation of the ultrastructure of a rock from which ultrafacies conclusions can be derived (usually diagenetic). The normal light microscope is usually sufficient for microfacies description (diagenetic or depositional), whereas core and field investigations allow the designation of depositional facies and additional diagenetic observations.

Depositional Facies

Depositional facies reflect a variety of chemical, mechanical and biological processes active in the depositional environment. They are referred to as *lithofacies* and/or *biofacies*. By reflecting environmental processes they also reflect aspects of depositional environments. Thus proper facies definition aids in recognising the depositional environment and provides a basis for extrapolation (see Fig. 2–3) over the zone in which depositional conditions are assumed to be constant. As such, facies definition on a regional scale is significant in oil geology. Often, informal terms such as *lithotype* or *rock type* replace the word *facies*. Such units are even occasionally merged into *rock units*. These terms and their usage coincide with the term *depositional facies* used here.

Lithofacies. Carbonate sedimentary facies are defined in terms of mineralogy, grain abundance and grain type, structures and textures all of which are described in Chapter 1. These combined parameters aid in reconstructing the depositional environment. When only some of these parameters are available, a full environmental reconstruction often cannot be made. This is apparent from the terms introduced by Folk (1962) (see Figs. 1–10 and 1–11). While it is true that a term such as *biopelmicrudite* gives a comprehensive summary of a number of important environmental parameters (grain type, grain size, depositional texture), it fails to denote such features as depositional structures, lateral consistency and mineralogy. Consequently Folk's terms are not facies terms and are often incomplete microfacies terms.

Biofacies. Skeletons of organisms are the most significant sediment contributors in carbonates. They give information about prevailing ecological conditions. Organisms sensitive to changes in environmental conditions (stenobiontic organisms) are most useful for environmental reconstruction. Organisms affect carbonate sediments in many ways other than by providing their skeletons (see Fig. 1–4).

Carbonate particles are not given to considerable transport owing to their easy wearability and solubility. Consequently organic skeletons in carbonate sediments usually reflect quite closely the life assembly (biocoenosis) of benthonic organisms, from which environmental parameters can be deduced. Thus the study of life assemblages forms the basis of biofacies recognition, and in carbonate sediments, in particular, it aids considerably in reconstructing the environment of deposition. A series of sketches of characteristic carbonate communities has been collected by McKerrow (1978), and the examples shown in Figs. 2–8 to 2–21 are taken from his work.

Depositional Facies in Time and Space

When regionally varying depositional conditions are pictured on a map showing the palaeogeography at an instant in time, the lateral facies distribution can be characterised either by regular facies zones or by a patchwork-quilt-like facies arrangement. If such environmental situations remain stationary over a time increment, a map representing that time interval can be sketched. However, environmental conditions usually change gradually over time increments, which results in facies shifting in place through time. If this has to be pictured, then a ratio map must be constructed, showing the average position of facies over a time increment. Environmental changes can be regular (cyclic) or irregular; they can affect areally restricted zones or extensive areas. Whether regular zonal or patchwork quilt facies distribution occurs in an area depends on the nature of the pre-existing topography and on the duration of the time increment over which environmental conditions remained stationary, allowing the development of mature carbonate facies. On a regular, very mildly sloping or a horizontal smooth shelf, broad facies zones are expected; on an irregular shelf topography, a patchwork quilt facies distribution is normal and on a steeply sloping shelf and shelf margin, narrow facies zones are common. Facies shifts in time result in vertically characteristic associations and sequences (Figs. 2–64 and 2–65), and if such sequences are suitably divided, they can be pictured laterally in averaged map distributions (Reijers, 1980; Burchette, 1981).

Proper recognition of facies sequences, facies associations and facies map pictures aids in designing appropriate sedimentation models that allow the extrapolation of particular facies into unknown areas (Reijers, 1972; Reijers and Ten Have, 1983; Bartok *et al.*, 1981). The basic law that allows the designation of sedimentation models has been formulated by Walther (1894). It states the following. Facies that were formed beside each other in space are superposed in section. When this law is applied to model designation, care should be taken that an uninterrupted succession is considered. Major breaks may represent the passage of any number of environments whose facies were removed.

When sedimentation models are based on facies sequences or cycles, the nature of the facies contacts has to be studied. Sharp vertical contacts are indicative of abrupt lateral changes, and sediments may represent environments which are widely separated in space. Only gradual, interfingering facies contacts are indicative of lateral continuation of a facies spectrum (Reijers, 1972).

Facies associations are groups of facies that are considered to reflect collectively an environmental setting or a genetic mechanism. Based on the configuration of the environmental setting, three groups of facies were recognised by Wilson (1975), who followed Rich (1951) and termed them facies associations (or facies belts) of the shelf (undaform), slope (clinoform) or basin plane (fondoform). Wilson worked this out in greater detail and arrived at nine standard facies (Fig. 2–74).

Lateral facies associations are influenced in time by relative sea level changes, giving rise to vertical facies associations of a transgressive, regressive or stationary nature. These are also referred to as build-up, build-out and build-in carbonate bodies (Reijers, 1980) or as hiatuses, depending on the sedimentation or erosional aspect of the environment. The mechanism behind the formation of such facies associations (Reijers, 1980; Longman, 1981) is equilibrium between relative sea level movements and carbonate production potential. This balance determines the chances of preservation of carbonate facies.

Diagenetic Facies

Diagenetic facies reflect a variety of processes that are active in diagenetic environments. In contrast to depositional facies, diagenetic facies also reflect the time increment during which diagenetic processes were active. The time effect modifies the nature of the early diagenetic fabrics and, while it is true that a diagenetic facies reflects a diagenetic environment at a certain moment, it is also true that ongoing accumulation of overburden results in early or shallow burial diagenetic facies moving into deeper burial diagenetic environments in which other characteristic processes act and leave their imprints on the fabric of the rock. Such overprints do not necessarily erase the previous diagenetic imprints. Aspects of diagenesis reflecting this thinking are discussed in Chapter 3.

Examples

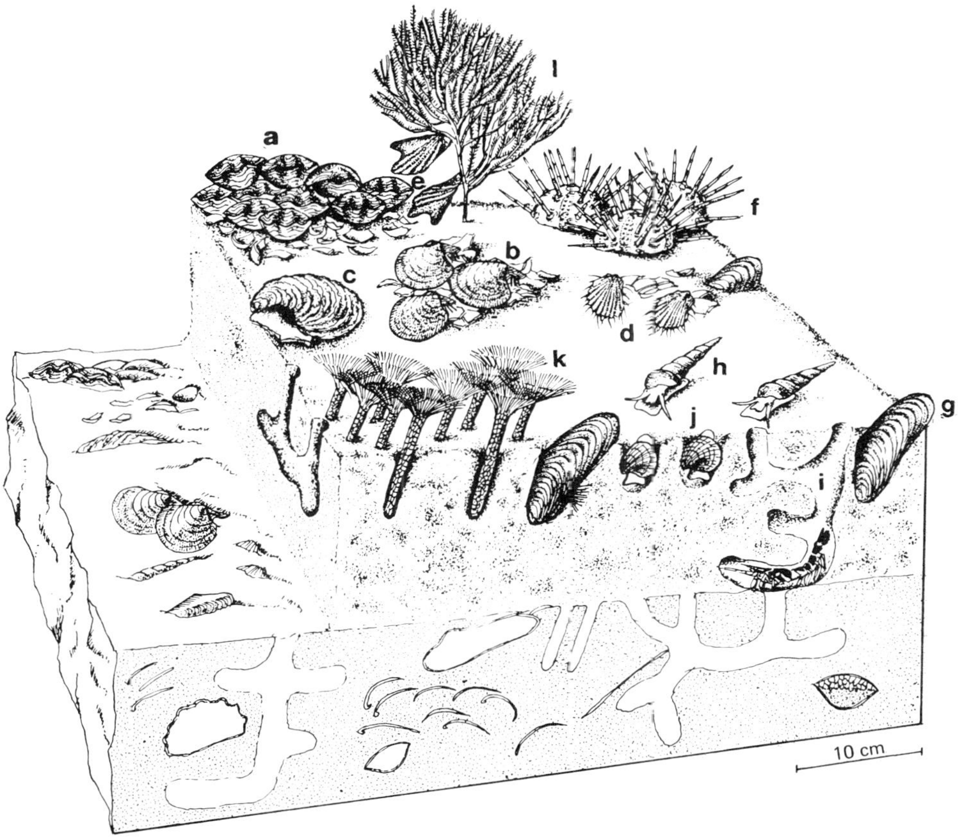

Fig. 2–8. Shelly lime mud community (Jurassic), which produces coquinoid limestone facies. (a) *Epithyris* (Brachiopoda: Articulata: Terebratulida), (b) *Camptonectes* (Mollusca: Bivalvia: Pterioida–pectinid), (c) *Liostrea* (Mollucsa: Bivalvia: Pterioida–oyster), (d) *Pseudolimea* (Mullusca: Bivalvia: Pterioidia), (e) *Costigervillia* (Mollusca: Bivalvia: Pterioida), (f) *Acrosalenia* (Echinodermata: Echinozoa), (g) *Modiolus* (Mollusca: Bivalvia: Mytiloida), (h) *Fibula* (Mollusca: Gastropoda: Archaeogastropoda), (i) *Thalassinoides* (trace-fossil) with burrowing *Glyphaea* (Crustacea), (j) *Anisocardia* (Mollusca: Bivalvia: Veneroida), (k) terebellid worms (Annelida), and (l) gorgonian (Coelenterata: Octocorallia) hypothetical in this reconstruction. After McKerrow (1979) (Fig. 79). By permission of the MIT Press.

Fig. 2–9. Santonian micraster chalk community (Cretaceous), which produces a chalky biomicrite facies. (a) *Parapuzosia* (Mollusca: Cephalopoda: Ammonoidea), (b) *Ventriculites* (Porifera: Hyalospongea), (c) *Echinocorys scutata* (Echinodermata: Echinoidea), (d) *Micraster* (Echinodermata: Echinoidea), (e) *Conulus albogalerus* (Echinodermata: Echinoidea), (f) *Bourgueticrinus ellipticus* (Echinodermata: Crinozoa–crinoid), (g) *Mitraster hunteri* (Echinodermata: Asterozoa–asteroid), (h) *Spondylus spinosus* (Mollusca: Bivalvia: Pterioida–pectinid), (i) *Inoceramus* fragments (Mollusca: Bivalvia: Pterioida), (j) *Pycnodonte vesicularis* (Mollusca: Bivalvia: Pterioida–oyster), (k) *Parasmilia centralis* (Coelenterata: Anthozoa: Scleractinia), (l) serpulids (Annelida), (m) bryozoans (Bryozoa: Ectoprocta), (n) *Gibbithyris* (Brachiopoda: Articulata: Terebratulida), (o) *Cretirhynchia* (Brachiopoda: Articulata: Rhynchonellida), (p) *Ptychodus* tooth (Vertebrata: Chondrichthyes), and (q) *Tylocidaris clavigera* (Echinodermata: Echinoidea). After McKerrow (1978) (Fig. 106). By permission of the MIT Press.

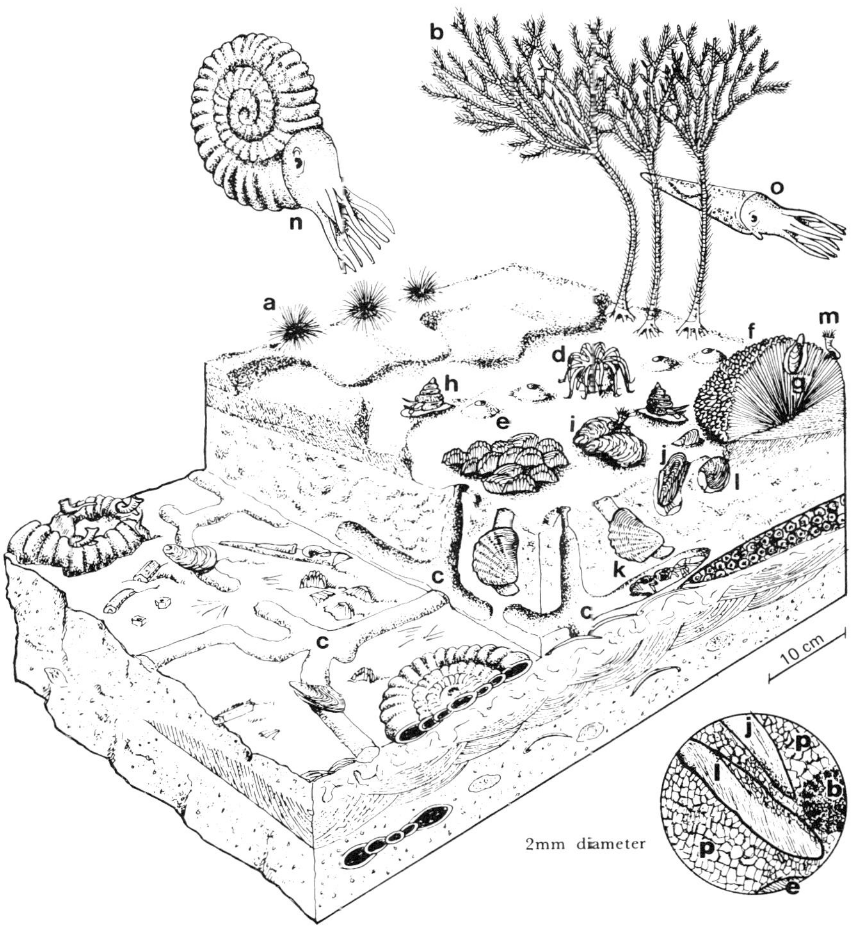

Fig. 2–10. Calcarenite community (Jurassic), which produces microcoquina–calcarenite facies. (a) Pseudodiadematid (Ectinodermata: Echinozoa–sea urchins), (b) *Pentacrinus* (Echinodermata: Crinozoa), (c) *Thalassinoides* (trace-fossil–crustacean), (d) *Oppelismilia* (Coelenterata: Anthozoa: Scleractinia), (e) rhynchonellids (Brachiopoda: Articulata: Rhynchonellida), (f) *Isastrea* (Coelenterata: Anthozoa: Scleractinia), (g) *Lithophaga* (Mollusca: Bivalvia: Mytiloida), (h) *Pleurotomaria* (Mollusca: Gastropoda: Archaeogastropoda), (i) *Liostrea* (Mollusca: Bivalvia: Pterioida–oysters), (j) *Modiolus* (Mollusca: Bivalvia: Mytiloida), (k) *Pholadomya* (Mollusca: Bivalvia: Anomalodesmata), (l) *Astarte* (Mollusca: Bivalvia: Veneroida), (m) serpulid (Annelida-polychaeta), (n) Uptonia (Mollusca: Cephalopoda: Ammoniodea), (o) belemnite (Mollusca: Cephalopoda: Coleoidea), and (p) cement of calcite. After McKerrow (1978) (Fig. 70). By permission of the MIT Press.

Fig. 2–11. Calcarenite community (Jurassic), which produces microcoquina/coquinite calcarenite facies. (a) *Rhaxella* (Porifera: Demospongea), (b) *Isognomon* (Mollusca: Bivalvia: Pterioida), (c) *Trigonia* (Mollusca: Bivalvia: Trigonioida), (d) *Camptonectes* (Mollusca: Bivalvia: Pterioida–pectinid), (e) serpulids (Annelida), (f) conjectural algal fronds, and (g) *Arenicolites* (trace-fossil of annelid). After McKerrow (1978) (Fig. 85). By permission of the MIT Press.

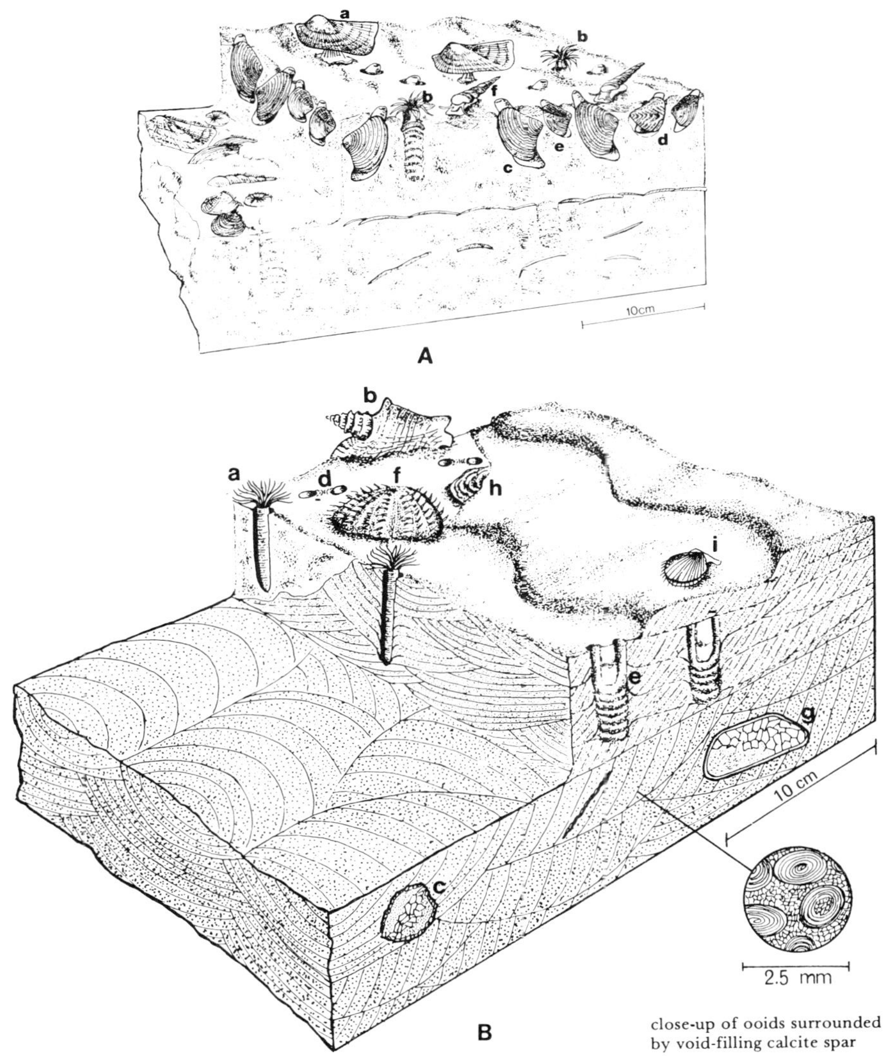

Fig. 2–12. A, Low-density, temporarily stable calcarenite community (Jurassic). Stationary sedimentation (reduced carbonate production), which produces calcarenite facies. (a) *Parallelodon* (Mollusca: Bivalvia: Arcoida), (b) *Chromatoseris* (Coelenterata: Anthozoa: Scleractinia), (c) *Eocallista* (Mollusca: Bivalvia: Veneroida), (d) *Vaughonia* (Mollusca: Bivalvia: Trigonioida), (e) Protocardia (Mollusca: Bivalvia: Veneroida), and (f) merineid gastropod (Mollusca: Gastropoda: Mesogastropoda). After McKerrow (1978) (Fig. 77). By permission of the MIT Press. B, Oolitic limestone community (Jurassic). High- to moderate-energy conditions and restricted fauna, which produce an oolite facies. (a) *Tigillites* (trace-fossil–annelid), (b) *Purpuroidea* (Mollusca: Gastropoda: Mesogastropoda), (c) shell of *Purpuroidea* replaced by calcite (Mollusca: Gastropoda), (d) opening of *Diplocraterion* (trace-fossil), (e) *Diplocraterion* showing response to oolite deposition, (f) *Pygaster* (Echinodermata: Echinozoa), (g) buried *Pygaster* part-filled with sediment and sparry calcite, (h) oyster fragment, and (i) pectinid (Mollusca: Bivalvia: Pterioida). After McKerrow (1978) (Fig. 76). By permission of the MIT Press.

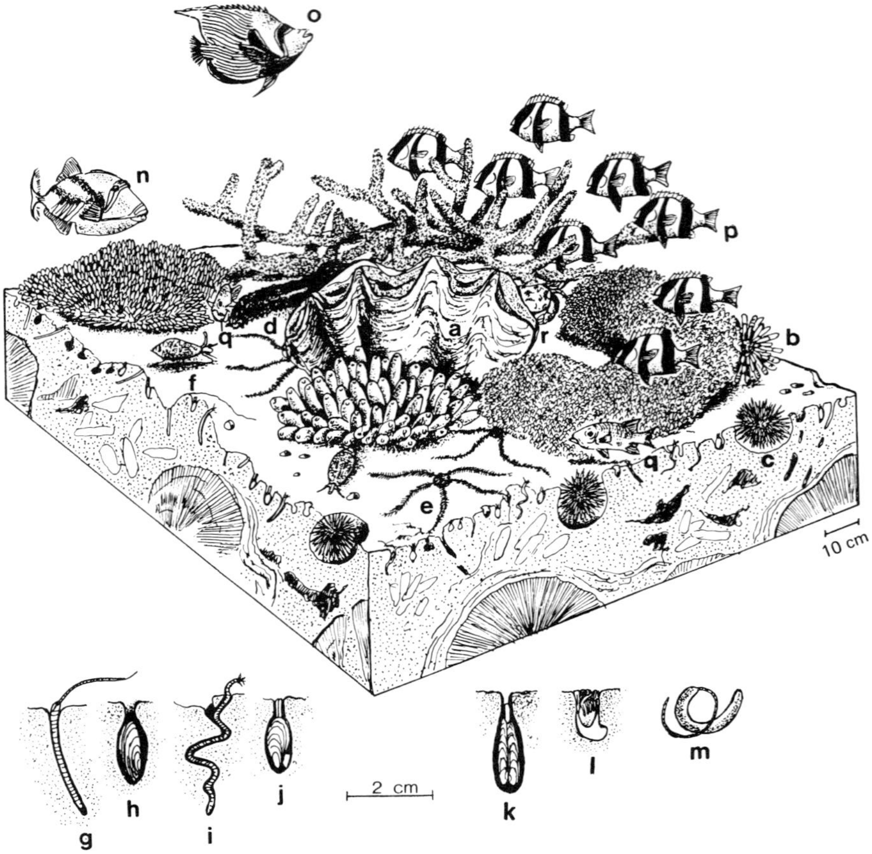

Fig. 2–13. Tropical coral reef assemblage (Recent), which produces reef/reefoid facies. (A) *Tridacna gigas* (Mollusca: Bivalvia: Veneroida), (b) *Heterocentrotus mamillathus* (Echinodermata: Echinoidea), (c) *Echinometra* (Echinodermata: Echinoidea), (d) *Stichopus chloronotus* (Echinodermata: Echinoidea), (e) *Ophiocoma* (Echinodermata: Asterozoa), (f) *Conus textile* (Mollusca; Gastropoda: Mesogastropoda), (g) *Aspidosiphon* (Sipunculida), (h) *Lithophaga* (Mollusca: Bivalvia: Mytiloida), (i) eunicid and nereid worms (Annelida), (j) *Rocellaria lamellosa* (Mollusca: Bivalvia: Myoida), (k) *Botula silicula* (Mollusca: Bivalvia: Mytiloida), (l) *Lithotrya* (Anthropoda: Crustacea: Cirripedia), (m) *Phascolosoma* (Sipunculida), (n) *Balistes* (Vertebrata: Osteichthyes–trigger fish), (o) *Pomacanthus* (Vertebrata: Osteichthyes), (p) *Dascyllus* (Vertebrata: Osteichthyes), (q) *Holocentrus spinifer* (Vertebrata: Osteichthyes), and (r) *Atergatis floridus* (Anthropoda: Crustacea: Malacostraca). After McKerrow (1978) (Fig. 125). By permission of the MIT Press.

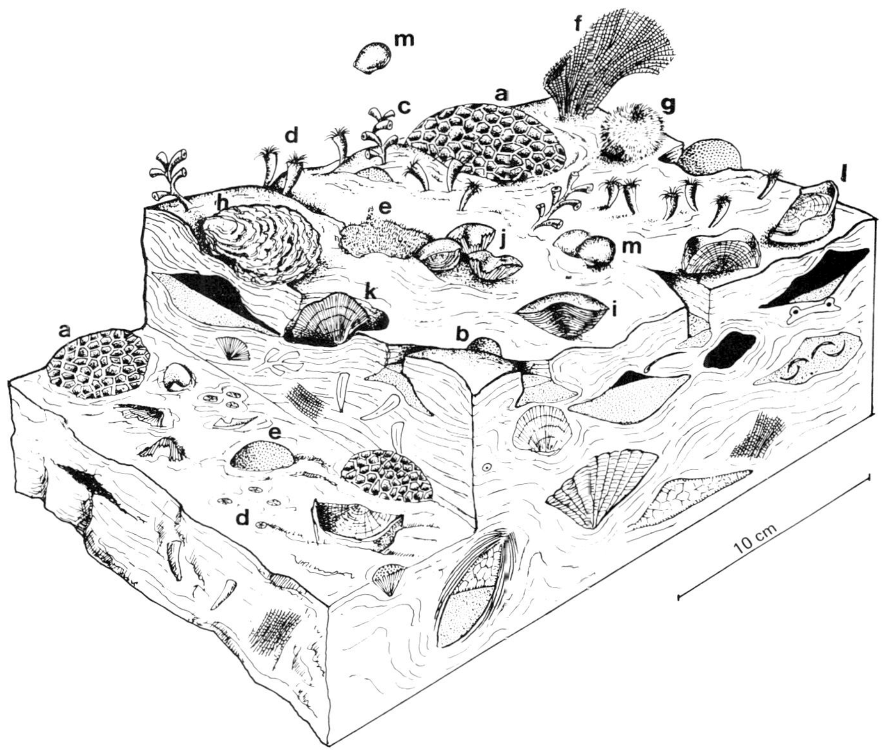

Fig. 2–14. Algal reef community (Carboniferous), which produces algal "reef" facies. (a) *Michelinia* (Coelenterata: Anthozoa: Tabulata), (b) *Emmonsia parasitica* (Coelenterata: Anthozoa: Tabulata), (c) *Cladochonus* (Coelenterata: Anthozoa: Tabulata), (d) *Cyathaxonia* (Coelenterata: Anthozoa: Rugosa), (e) *Fistulipora* (Bryozoa: Ectoprocta), (f) *Fenestella* (Bryozoa: Ectoprocta), (g) lithistid sponge (Porifera: Demospongea) (h) *Pachypteria* (Mollusca: Bivalvia: Pterioida), (i) '*Reticularia*' (Brachiopod: Articulata: Spiriferida), (j) *Stenoscisma* (Brachiopod: Articulata: Rhynchonellida), (k) *Streptorhynchus* (Brachiopod: Articulata: Strophomenida), (l) *Leptagonia* (Brachiopod: Articulata: Strophomenida), and (m) *Entomoconchus* (Anthropoda: Crustacea: Ostracoda). After McKerrow (1978) (Fig. 44). By permission of the MIT Press.

Fig. 2–15. A, Barrier reef top community (Permian) with abundant Bryozoan, Brachiopodal and coral growth, which produces reef flat facies. (a) *Acanthocladia anceps* (Bryozoa: Ectoprocta), (b) *Fenestella retiformis* (Bryozoa: Ectoprocta), (c) *Synocladia virgulacea* (Bryozoa: Ectoprocta), (d) *Thamniscus dubius* (Bryozoa: Ectoprocta), (e) *Dielasma elongatum* (Brachiopoda: Articulata: Terebratulida), (f) *Horridonia horrida* (Brachiopoda: Articulata: Strophomenida), (g) *Pterospirifer alatus* (Brachiopoda: Articulata: Spiriferida), (h) *Stenoscisma humbletonensis* (Brachiopoda: Articulata: Rhynchonellida), (i) *Mourlonia ? linkiana* (Mollusca: Gastropoda: Archaeogastropoda), (j) *Bakevellia binneyi* (Mollusca: Bivalvia: Pterioida), (k) *Parallelodon striatus* (Mollusca: Bivalvia: Arcoida), and (l) *Streblochondria? pusilla* (Mollusca: Bivalvia: Pterioida). After McKerrow (1978) (Fig. 57). By permission of the MIT Press. B, Upper reef slope community (Carboniferous) with abundant coral, Brachiopodal and Lamelli branchiata growth, which produces reef-slope facies. (a) *Caninia* (Coelenterata: Anthozoa: Rugosa), (b) *Hexaphyllia* (Coelenterata: Anthozoa: Heterocorallia), (c) *Linoprotonia* (Brachiopoda: Articulata: Strophomenida), (d) *Echinoconchus elegans* (Brachiopoda: Articulata: Strophomenida), (e) *Acanthoplecta mesoloba* (Brachiopoda: Articulata: Strophomenida), (f) *Productina margaritaceus* (Brachiopoda: Articulata: Strophomenida), (g) *Alitaria* (Brachiopoda: Articulata: Strophomenida), (h) *Rhipidomella* (Brachiopoda: Articulata: Orthida), (i) *Aulacophoria keyserlingiana* (Brachiopoda: Articulata: Orthida), (j) *Spiriferellina insculpta* (Brachiopoda: Articulata: Spiriferida), (k) *Dielasma* (Brachiopoda: Articulata: Terebratulida, (l)

Fig. 2–15. (*Continued*)

Pleuropugnoides pleurodon (Brachiopoda: Articulata: Rhynchonellida), (m) *Conocardium* (Mollusca: Rostroconchia), *Parallelodon* (Mollusca: Bivalvia: Arcoida), (o) *Straparollus* (Mollusca: Gastropoda: Archaeogastropoda), (p) *Bellerophon* (Mollusca: Monoplacophora), (q) *Cummingella* (Arthropoda: Trilibita), (r) *Rhombocladia* (Bryozoa: Ectoprocta), (s) *Penniretepora* (Bryozoa: Ectoprocta), (t) Koninckopora (Algae), (u) algal encrusted shells, and (v) seaweed (Algae). After McKerrow (1978) (Fig. 45). By permission of the MIT Press.

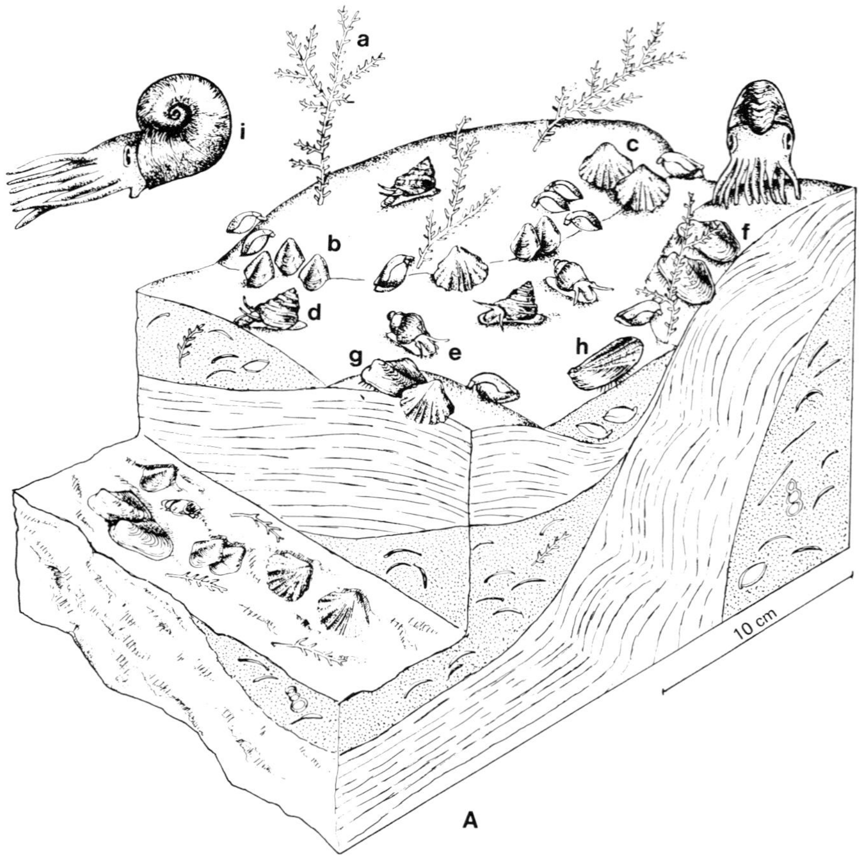

Fig. 2–16. A, Backreef community (Permian) with abundant restricted marine faunal elements, which produces backreef facies. (a) *Acanthocladia anceps* (Bryozoa: Ectoprocta), (b) *Dielasma elongatum* (Brachiopoda: Articulata: Terebratulida), (c) *Stenoscisma schlotheimi* (Brachiopoda: Articulata: Rhynchonellida), (d) *Yunnania* (Mollusca: Gastropoda, Archaeogastropoda), (e) *Naticopsis minima* (Mollusca: Gastropoda: Archaeogastropoda), (f) *Bakevellia binneyi* (Mollusca: Bivalvia: Pterioida), (g) *B. ceratophaga* (Mollusca: Bivalvia: Pterioida), (h) *Permophorus costatus* (Mollusca: Bivalvia: Veneroida), and (i) *Peripetoceras freieslebeni* (Mollusca: Cephalopoda: Nautiloidea). After McKerrow (1978) (Fig. 58). By permission of the MIT Press. B, Silurian reef assemblage. (a) *Heliolites* (Coelenterata: Anthozoa: Tabulata), (b) *Favosites* (Coelenterata: Anthozoa: Tabulata), (c) *Halysites* (Coelenterata: Anthozoa: Tabulata), (d) *Hallopora* (Bryozoa: Ectoprocta), (e) streptelasmatid (Coelenterata: Anthozoa: Rugosa), (f) *Atrypa* (Brachiopoda: Spiriferida), (g) crinoid (Echinodermata: Crinozoa), (h) *Leptaena* (Brachiopoda: Strophomenida), (i) *Dalmanites* (Anthropoda: Trilobita), and (j) orthocone (Mollusca: Cephalopoda: Nautiloidea). After McKerrow (1978) (Fig. 24). By permission of the MIT Press.

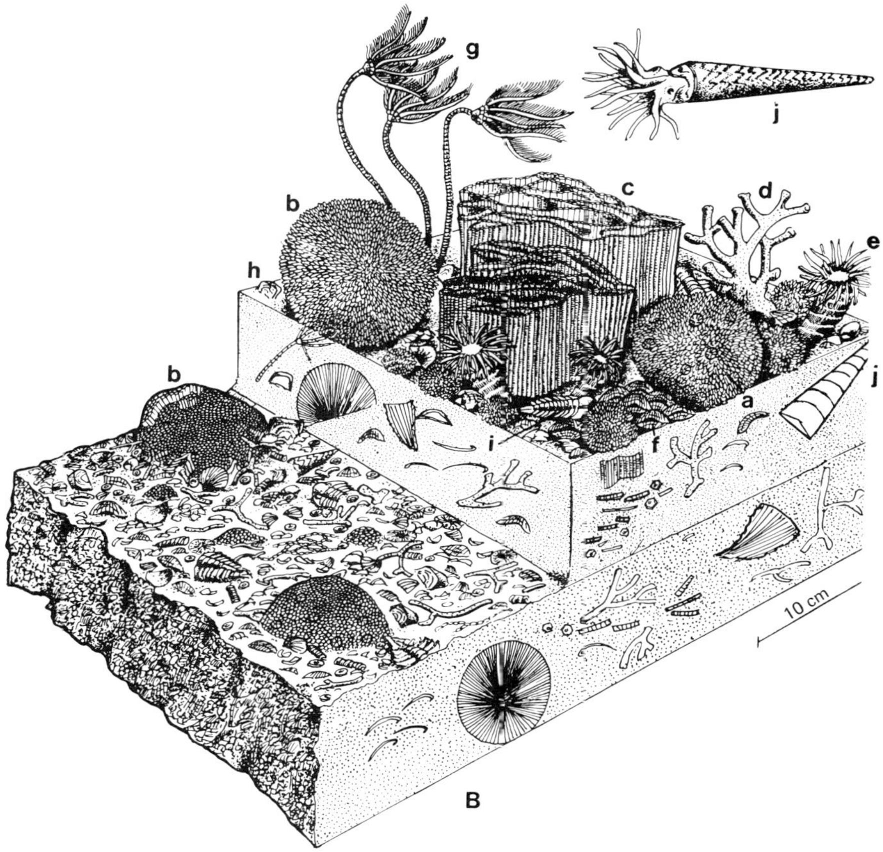

Fig. 2–16. (*Continued*)

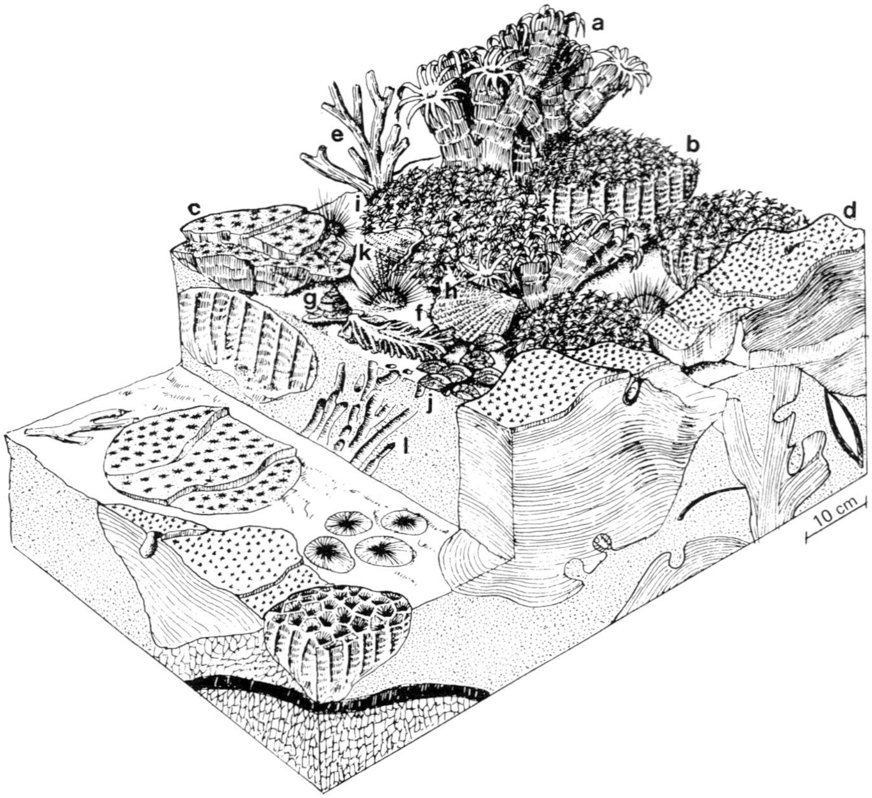

Fig. 2–17. Coral–algal patch-reef community (Jurassic) with abundant massive platy coral forms, which produces patch reef facies. (a) *Thecosmilia* (Coelenterata: Anthozoa: Scleratinia), (b) *Isastrea* (Coelenterata: Anthozoa: Scleractinia), (c) *Thamnasteria arachnoides* (Coelenterata: Anthozoa: Scleractinia), (d) *Thamnasteria concinna* (Coelenterata: Anthozoa: Scleractinia), (e) *Rhabdophyllia* (Coelenterata: Anthozoa: Scleractinia), (f) *Lopha* (Mollusca: Bivalvia: Pterioida), (g) trochid (Mollusca: Gastropoda: Archaeogastropoda), (h) *Chlamys* (Mollusca: Bivalvia: Pterioida–pectinid), (i) *Cidaris* (Echinodermata: Echinozoa–sea urchin), (j) terebratulids (Brachiopoda: Articulata: Terebratulida), (k) bryozoan (Bryozoa: Ectoprocta), and (l) *Cladophyllia conybeari* (Coelenterata: Anthozoa: Scleractinia). After McKerrow (1978) (Fig. 87). By permission of the MIT Press.

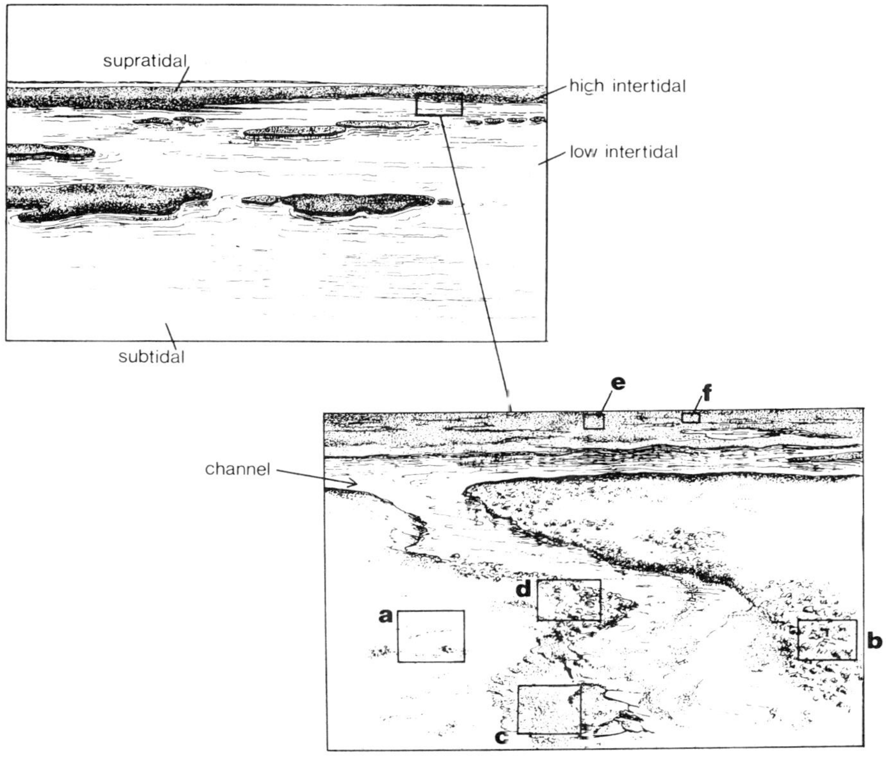

Fig. 2–18. A, View from the sea of a sabkha coast near Abu Dhabi with details on the channel structures. Tracings of photographs. B, Channel (a few metres wide) on a sabkha coast (Jurassic). Locations a–f correspond to insets a–f of A. (a) Entrance to crab burrow—trace of *Callianassa* (Crustacea), (b1) trails of grazing gastropods on an algal surface, (b2) entrances to 'ghost shrimp' burrows, (c) surface of crinkled algal mat; blue-green alga (Prokaryota–stromatolite), (d) surface covered with crustacean faecal pellets, (e1) pustular algal mat, (e2) algal mat cut by dessication cracks, and (f) stromatolitic laminations with gypsum pseudomorphs and vugs.

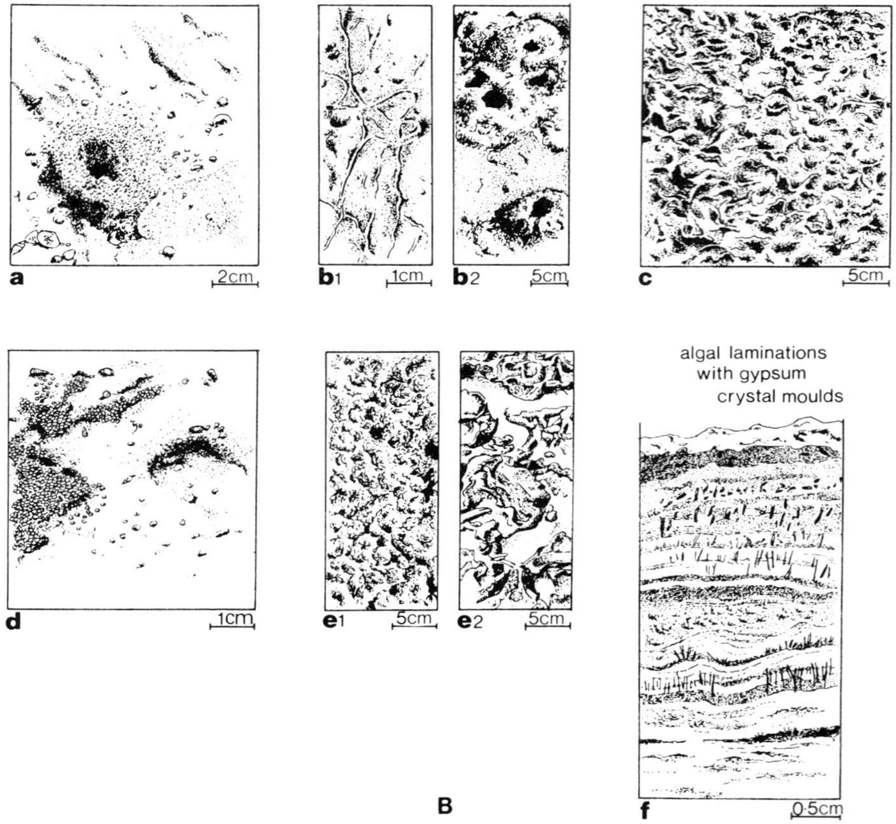

Fig. 2–18. (*Continued*)

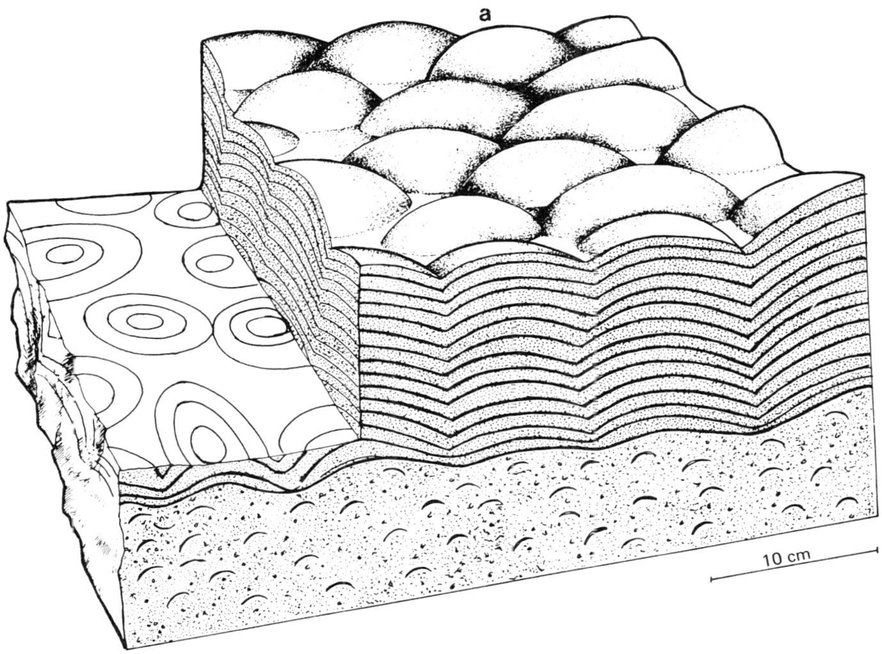

Fig. 2–19. Tidal flat communities (Carboniferous) with abundant algal growth, which produce tidal-flat facies. (a) Stromatolite domes (algae), (b) stromatolite pillars (Algae), (c) *Polidevcia attenuata* (Mollusca: Bivalvia: Palaeotaxodonta), (d) *Spirorbis* (Annelida: Serpulid), (e) temporary bridging laminae between pillars, (f) carbonate sand, (g) sediment-rich laminae (in inset), and (h) algal-rich laminae (in inset). After McKerrow (1978) (Figs. 39, 40). By permission of the MIT Press.

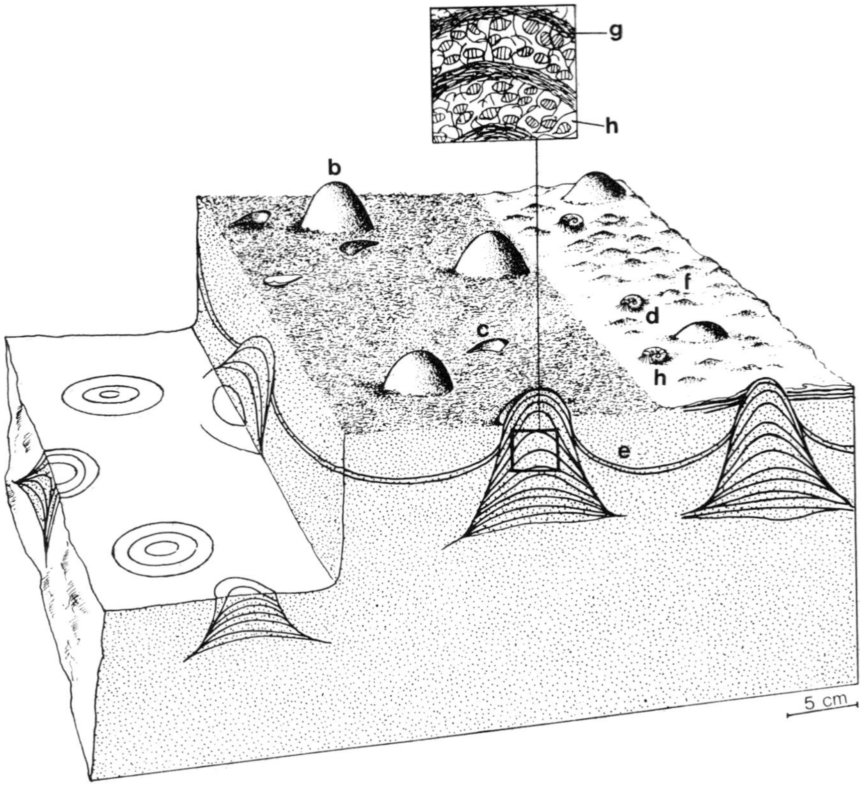

Fig. 2–19. (*Continued*)

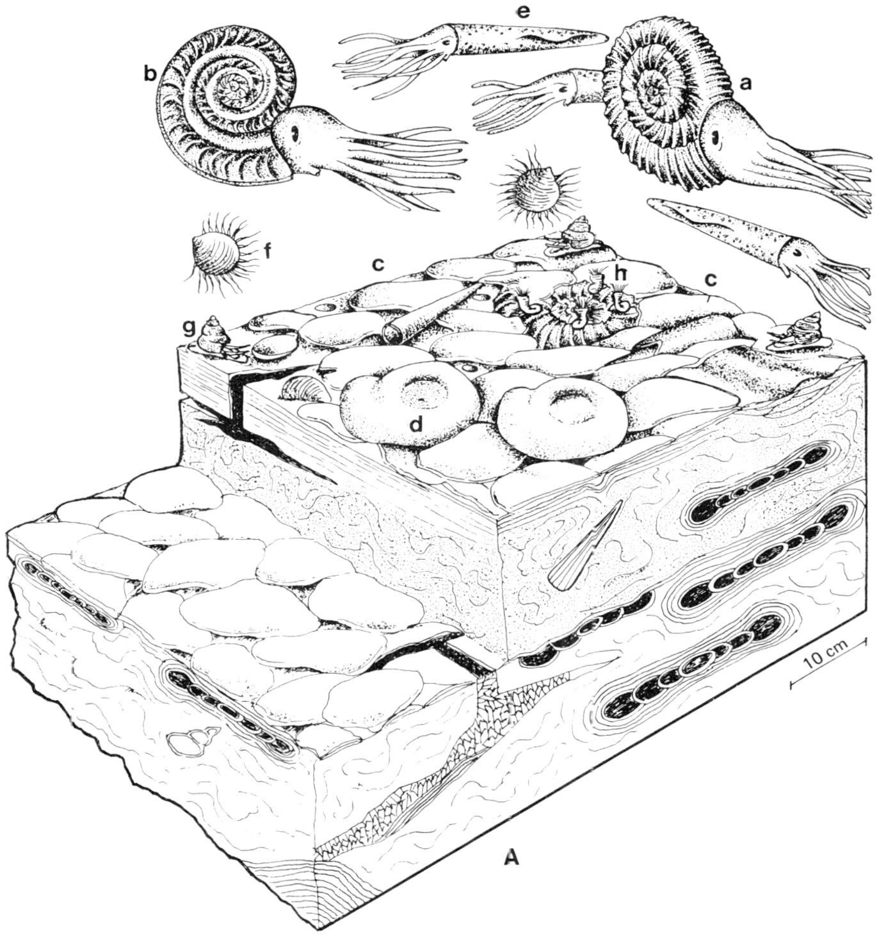

Fig. 2–20. A, Condensed limestone community (Jurassic), which produces biomicrite facies. (a) *Dactylioceras* (Mollusca: Cephalopoda: Ammnoidea), (b) *Hildoceras* (Mollusca: Cephalopoda: Ammonoidea), (c) algae (Prokaryota–oncolites), (d) ammonite covered with algae as in (c), (e) belemnite (Mollusca: Cephalopoda: Coleoidea), (f) *Eositra* (Mollusca: Bivalvia: Pterioida), (g) *Amberleya* (Mollusca: Gastropoda: Mesogastropoda), (h) serpulids (Annelida–polychaete). After McKerrow (1978) (Fig. 68). By permission of the MIT Press. B. Phosphatic nodule-bed community (Cretaceous). After McKerrow (1978) (Fig. 96). By permission of the MIT Press.

Fig. 2–20. (*Continued*)

Fig. 2–21. Hardground community (Jurassic) with no sedimentation and consolidation of sediment, which produces condensed sequence facies. (a) *Pentacrinus* (Echinodermata: Crinozoa), (b) *Liostrea* (Mollusca: Bivalvia: Pterioida–oyster), (c) *Montlivaltia* (Coelenterata: Anthozoa: Scleractinia), (d) echinoids (Echinodermata: Echinozoa–sea urchin), (e) serpulids (Annelida–polychaete), (f) boring worms (Annelida), (g) *Lithophaga* (Mullosca: Bivalvia: Mytiloida), (h) bryozoan (Bryozoa: Ectoprocta), (i) patellid (Mollusca: Gastropod: Archaeogastropod), (j) encrusting foraminiferida (Sarcodina), (k) *Amaltheus* (Mollusca: Cephalopod: Ammonoidea), (l) belemnite (Mollusca: Cephalopod: Coleoidea), (m) *Pseudopecten* (Mollusca: Bivalvia: Pterioida–pectinid), (n) *Tetrarhynchia* (Brachiopoda: Articulata: Rhynchonellida), (x) bedrock (pre-Jurassic), and (y) hardened Jurassic sediment. After McKerrow (1978) (Fig. 71). By permission of the MIT Press.

Definition 185

185. MICROFACIES: the total of the palaeontological and sedimentological criteria recognisable in thin sections, peels and polished slabs.

Note: See Flügel (1978, 1982).

NE microfacies
DE Mikrofazies
ES microfacies
FR microfaciès
RE —

2–4 CARBONATE DEPOSITIONAL SETTINGS AND BODIES (RECENT)

Introduction: General Terms

Carbonate depositional settings are the surroundings in which carbonates are produced or come to rest and form sediment accumulations. Such accumulations reflect the depositional environment (facies) and the setting. Carbonate bodies form after a long period of accumulation. The ultimate shape of the bodies is controlled by the following:

1. Carbonate production potential of the environment: the capacity of the environment to encourage the biochemical production of carbonates
2. Sedimentation potential of the environment: the capacity of the environment to encourage the transportation and/or the stabilisation/fixation of carbonates
3. Structural control: regional subsidence that might be slower, equal to or faster than carbonate production or sediment accumulation

Depositional settings are controlled by the configuration of the shoreline and that of the sea bottom. The climate to a certain degree controls the carbonate production potential and consequently the shape of the setting. The wind controls wave and current patterns and consequently to a certain degree the sedimentation potential.

Wilson (1975) recognised megatectonic provinces characterised by considerable and moderate subsidence. In such terms the following marine carbonate bodies and settings can be recognised:

1. Basinal build-up carbonate bodies: major offshore banks such as the Great Bahama Bank and isolated entities (usually smaller than major offshore banks) such as atolls
2. Shelf-margin build-out carbonate bodies: wide and large carbonate masses building out and building up into major platforms and ramps (e.g., at the trailing edge of a continental plate)
3. Shelf-margin build-in masses: narrow fringes or haloes of carbonate masses (e.g., at the leading edge of a continental plate)

4. Shelf-interior carbonate masses: comparable to shelf-margin build-outs at the trailing edge of a continental plate but smaller and irregular when the relief of the shelf interior is low

Moreover, build-in and build-up masses occur as a response to steady or episodic subsidence and/or sea level fluctuation. Thus the resulting shelf strata are frequently cyclic in nature. Depending on the degree of regional subsidence and the sedimentation potential, shelf-interior masses may shoal to such a degree that intertidal or supratidal (evaporitic) flats and/or sabkha settings are formed, each with characteristic sediment bodies. Reed (1985) gives a comprehensive overview of Recent and ancient carbonate platform facies models.

The arrangement of defined terms hereafter follows the outline sketched above. First coastal marine or terrestrial carbonate sedimentary bodies are defined. Next, terms related to shallow marine carbonate sedimentary bodies are discussed in a section forming the bulk of the chapter.

Definitions 186–191

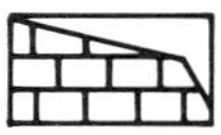

186. CARBONATE RAMP (FACIES): huge carbonate bodies built away from positive areas and down gentle regional palaeoslopes. No striking break in slope exists, and facies patterns are apt to be wide and regular. Belts with the highest energy zone are relatively close to the shore (see Fig. 2–22).

NE karbonaathelling
DE flach einfallende Karbonatplattform, Karbonatrampe
ES rampa carbonatado; facies carbonatadas de rampa
FR pente à sédimentation carbonatée

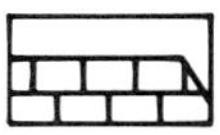

187. CARBONATE PLATFORM (FACIES): huge carbonate bodies built up with a more or less horizontal top and abrupt shelf margins where "high-energy" sediments occur (see Fig. 2–23).

Note: The normal process of carbonate sedimentation effectively and rapidly turns ramps into platforms and creates narrow, steep shelf-margin ridges. Slopes on some ramps may be so gentle as to be commonly indistinguishable from platforms. Thus these terms are often used interchangeably.

NE karbonaatvlakte; karbonaatplatform
DE Karbonatplattform

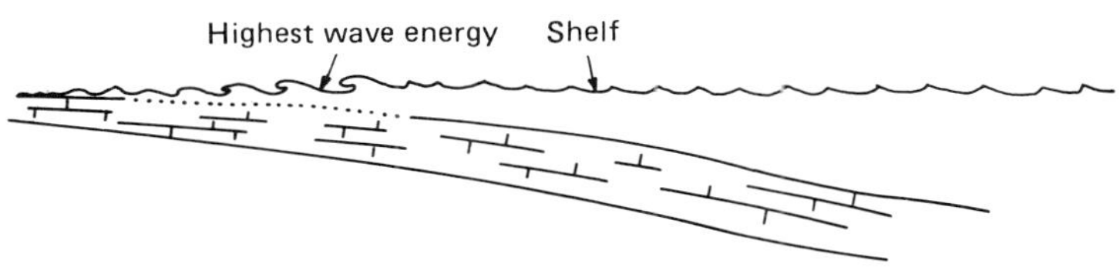

Fig. 2–22. Carbonate ramp as introduced by Ahr (1973) and further defined by Wilson (1975). Illustration by kind permission of Dr. J.L. Wilson and Springer Verlag.

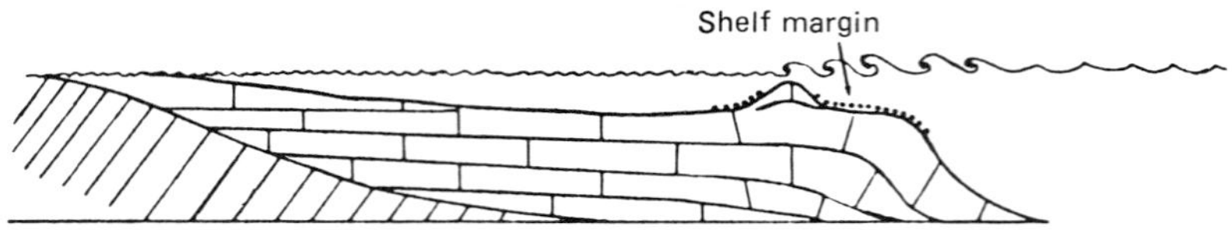

Fig. 2–23. Carbonate ramp (Definition 186) evolved into a carbonate platform at the margin of which local carbonate build-ups occur. Illustration by kind permission of Dr. J.L. Wilson and Springer Verlag.

ES plataforma carbonatada; facies carbonatades de plataforma
FR plateforme carbonatée

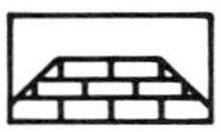

188. OFFSHORE CARBONATE BANK: complex carbonate build-up, often of great size and thickness (e.g., Bahama Bank), well offshore from the coastal carbonate ramps or carbonate platforms (see Fig. 2–24).

NE karbonaatbank in open zee
DE Karbonatbank im Schelfbereich
ES banco carbonatado situado mar adentro
FR plateforme carbonatée isolée (du continent), en haute mer
RE SENS

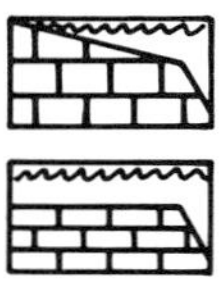

189. CARBONATE SHELF: a semiflat area on top of a carbonate ramp or a carbonate platform (see Fig. 2–25).

NE karbonaatvlakte; karbonaatplatform
DE Karbonatschelf
ES plataforma continental carbonatada
FR plateforme continentale carbonatée
RE MS

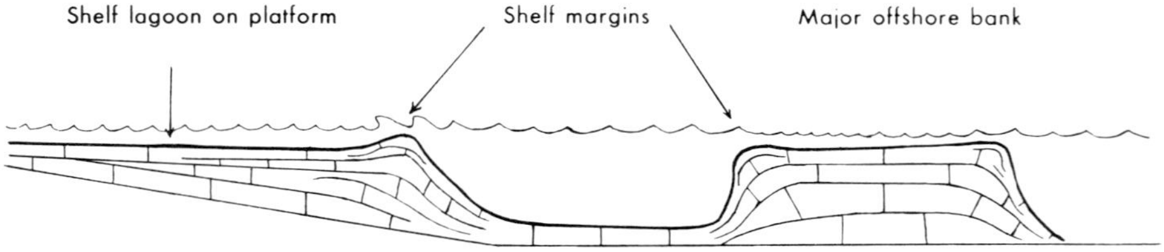

Fig. 2–24. Carbonate banks as opposed to carbonate platforms (or carbonate shelves). Note the development of shelf-margin facies in both carbonate banks and carbonate platforms.

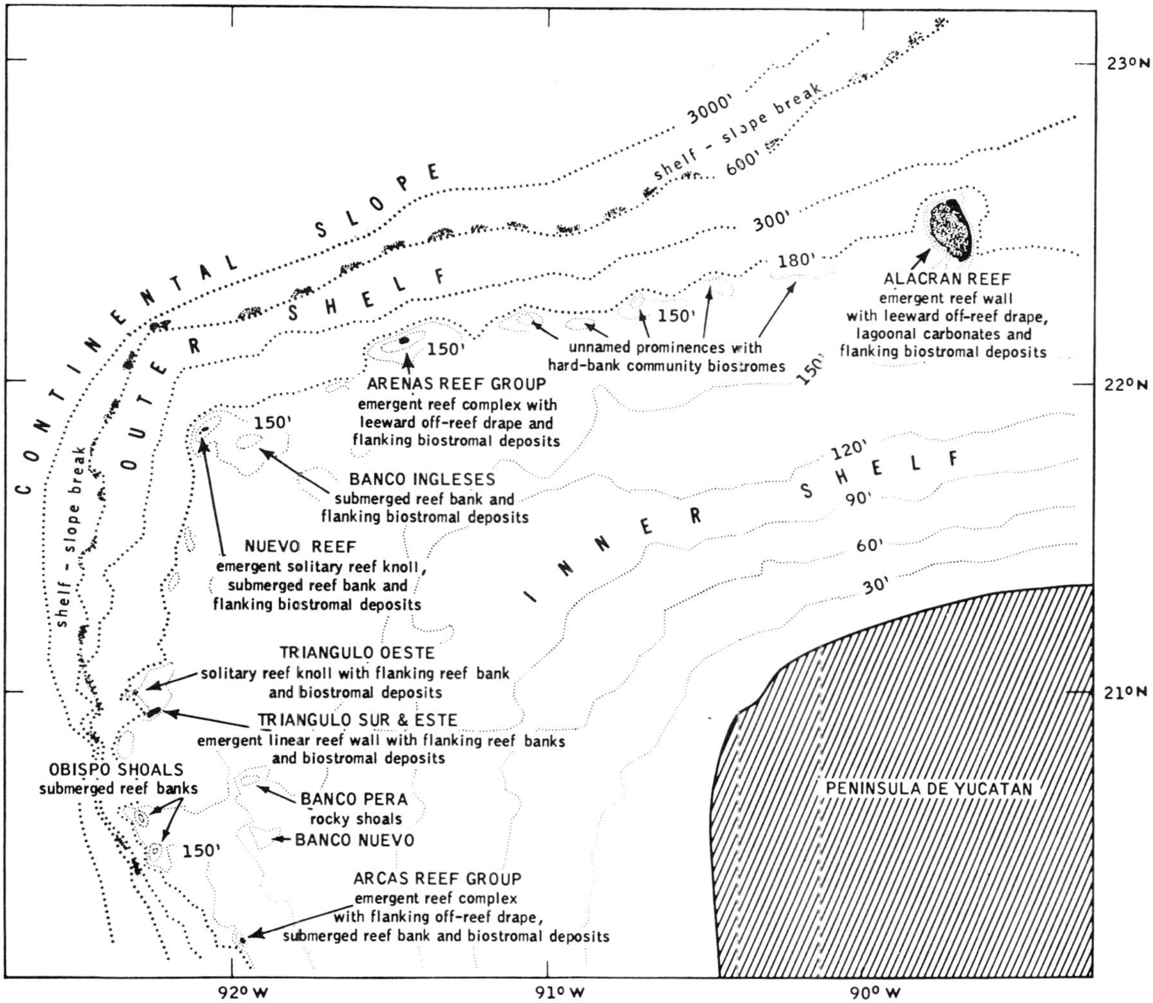

Fig. 2–25. Distribution of emergent reef masses, submerged reef banks and hard-bank biostromes in Yucatán shelf (Logan *et al.*, 1969). Illustration by permission of AAPG.

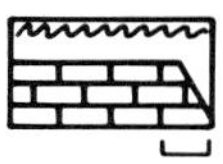

190. CARBONATE SHELF MARGIN: the edge of a carbonate shelf (see Fig. 2–26).

NE rand van een karbonaatvlakte
DE Karbonat Schelfrand; Schelfkante
ES borde del plataforma carbonatada
FR rebord de plateforme carbonatée
RE OS

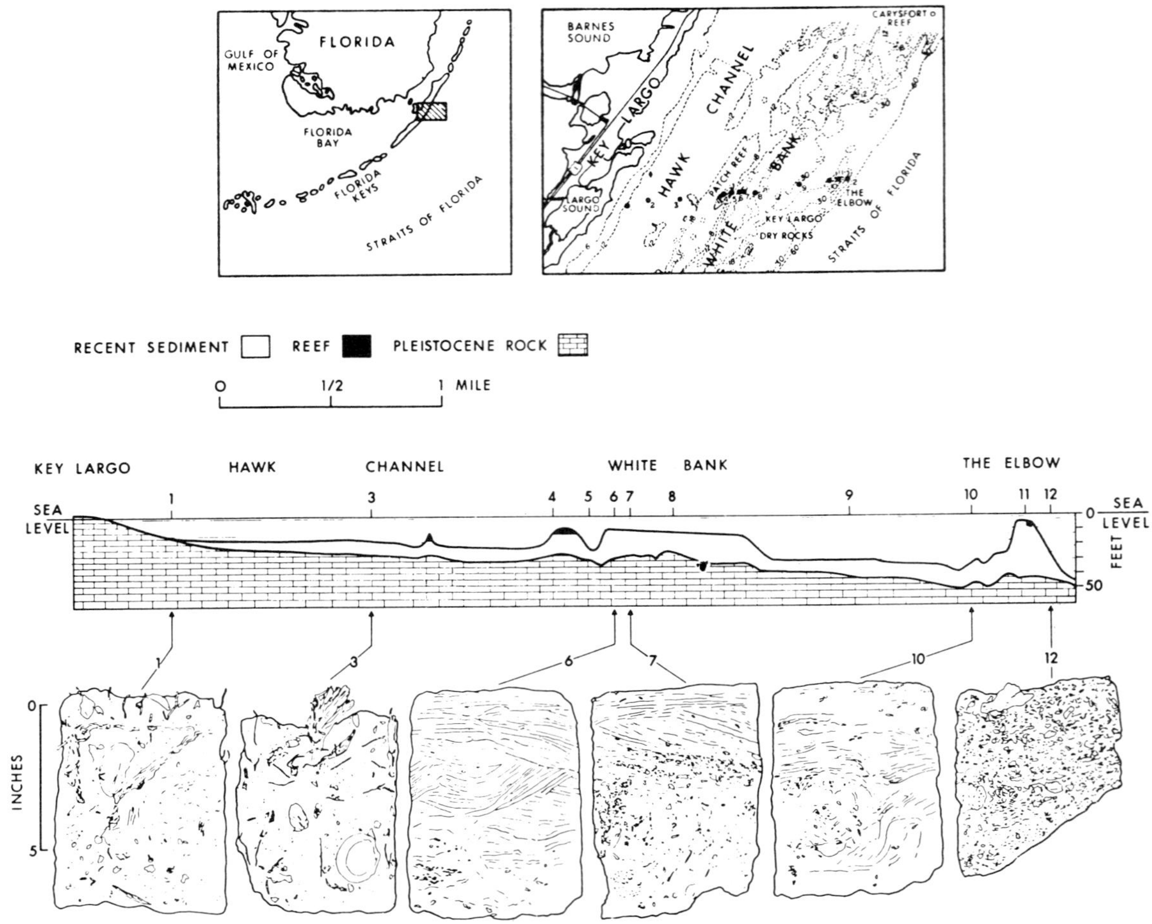

Fig. 2–26. Sedimentary structures of the shelf margin as traced from surface-impregnated box cores. Illustration courtesy of Geol. Soc. Am. and Dr. P. Enos.

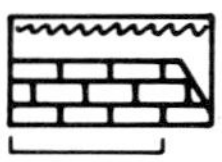

191. CARBONATE SHELF LAGOON: a term often used for the shallow shelf seas on a carbonate platform (see Fig. 2–27).

NE lagune op een karbonaatvlakte
DE Karbonatschelf Lagunen
ES laguna de plataforma carbonatada; ''lagoon'' de plataforma carbonatada
FR lagon de plateforme carbonatée
RE SENL

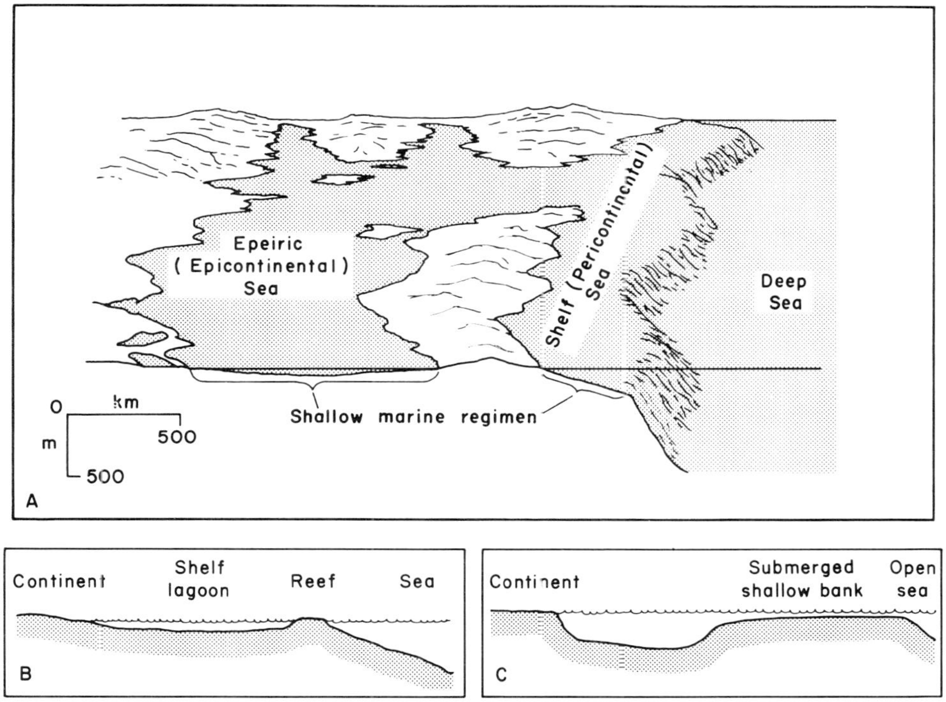

Fig. 2–27. Schematic views of shelf and epeiric seas. A, Pericontinental shelf and epeiric (epicontinental) sea. Shelf seas cover margin of modern continental blocks. Epeiric seas are not present in the Recent. In the past they spread over large parts of continents. B, Shelf lagoon occurs between margin of an oceanic block and reef tracts situated at the edge of the continental mass. C, Tidal sea covers a submerged shallow bank surrounded by deep sea. Illustration after Heckel (1972). By permission of Soc. Econ. Paleontol. Mineral

Tidal Carbonate Settings and Bodies

Recognition of carbonate depositional environments, settings and bodies in ancient rocks is considerably enhanced if a comprehensive data set is available for comparison with Recent carbonate depositional criteria. Thus it is not surprising that knowledge of tidal environments (the meaning of *tidal* in the broadest sense; e.g., short-period fluctuation in sea level due to astronomical or meteorological causes, or both) has advanced rapidly. Such environments are the most accessible.

The tidal flat is the most important tidal setting. It straddles the uncertain boundary between land and sea and is subject to the widest spectrum of environmental conditions. The reflections of these show an often bewildering variety of sedimentary features. Many of the structures do not reveal anything of consequence about the environmental conditions; it is in association that these features tell their story. Nor are many of the structures exclusively characteristic of carbonates;

siliciclastic tidal flats often show considerable structural or textural analogies. This section concentrates on significant carbonate structural or textural terms; those that have a limited significance or are not very characteristic of carbonates have been omitted). It is for these reasons that the following terms and their definitions are not exhaustive and that the choice is arbitrary and open to amendment. The terms are arranged in a sequence that goes from the offshore area towards the tidal flat proper. Some supratidal and terrestrial terms conclude the section.

Definitions 192–202

192. TIDAL FLAT BEACH RIDGE: a natural minor elevation just reaching above normal tide range and causing the development of protected channel–levee–pond complexes in the tidal flat province.

NE getijden strandwal
DE Wattstrandwall
ES cresta de llanura mareal
FR levée de baie; levée de slikke
RE SENN; CTN

193. BEACH RIDGE WASHOVER: lobes or fans of sediment that make wide, low saddles between the hummocks at the crest of the beach ridge.

NE strandwal overspoeling
DE Sturmdelta
ES delta de oleaje en un cordón de playa
FR delta par rupture de cordon littoral
RE SENF; CTF

194. BEACH RIDGE HUMMOCKS: sand mounds, elongated and parallel to the shoreline, that are overtopped by the sea only during hurricane surges. They form the highest positive features in the tidal flat province (see Fig. 2–28). Essentially they are a string of islands.

NE strandwal eilanden
DE Uferwall
ES cresta de playa
FR crête de plage; levée de plage
RE SENB; CTB

195. TIDAL BAR: an elongate sediment body formed under the influence of tidal currents.

NE getijdenstrandwal; getijdenbank
DE Gezeiten Barre; Gezeiten-Strandwall

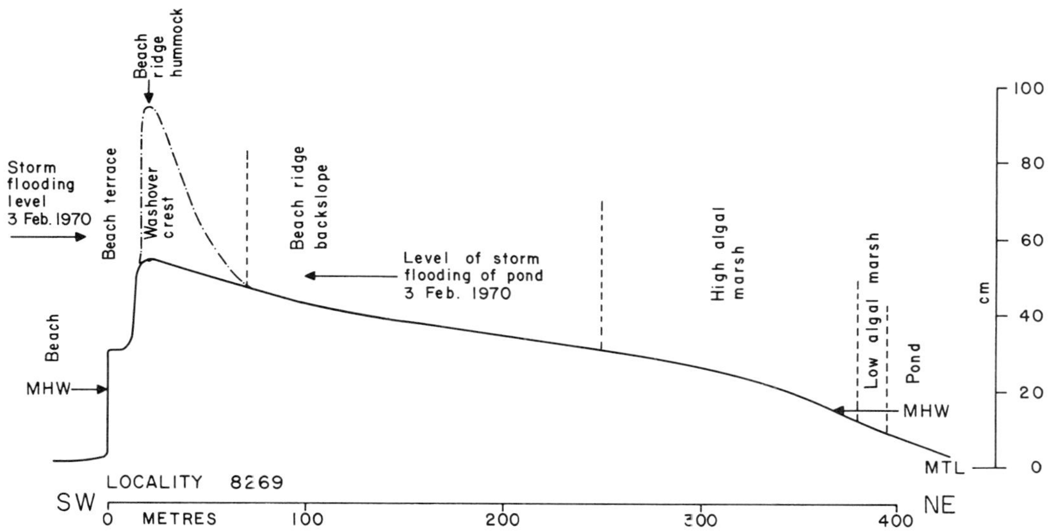

Fig. 2–28. Profile across a beach from beach ridge hummock to pond showing the vertical distribution of subenvironments and settings at the shoreline. Illustration after Hardie (1977). By permission of The John Hopkins University Press.

ES barra de marea
FR cordon de marée
RE SENB; CAB

196. TIDAL CHANNEL: sinuous, often branching channel through which the tide is brought into and out of the tidal flat province. Such channels are bordered by low natural levees behind which "ponds" may be present (see Fig. 2–29).

NE getijdenkanaal; getijdengeul
DE Wattrinne; Priel
ES canal de marea
FR chenal de marée
RE SENC; CAC

197. TIDAL FLAT: the setting in the supratidal, intertidal and shallow marine environments in which the tides are a decisive environmental parameter which is reflected in a wide variety of sedimentary and diagenetic textures and structures (see Fig. 2–30).

NE getijdenvlakte; wad
DE Watt; Wattfläche
ES llanura mareal; llanura de marea

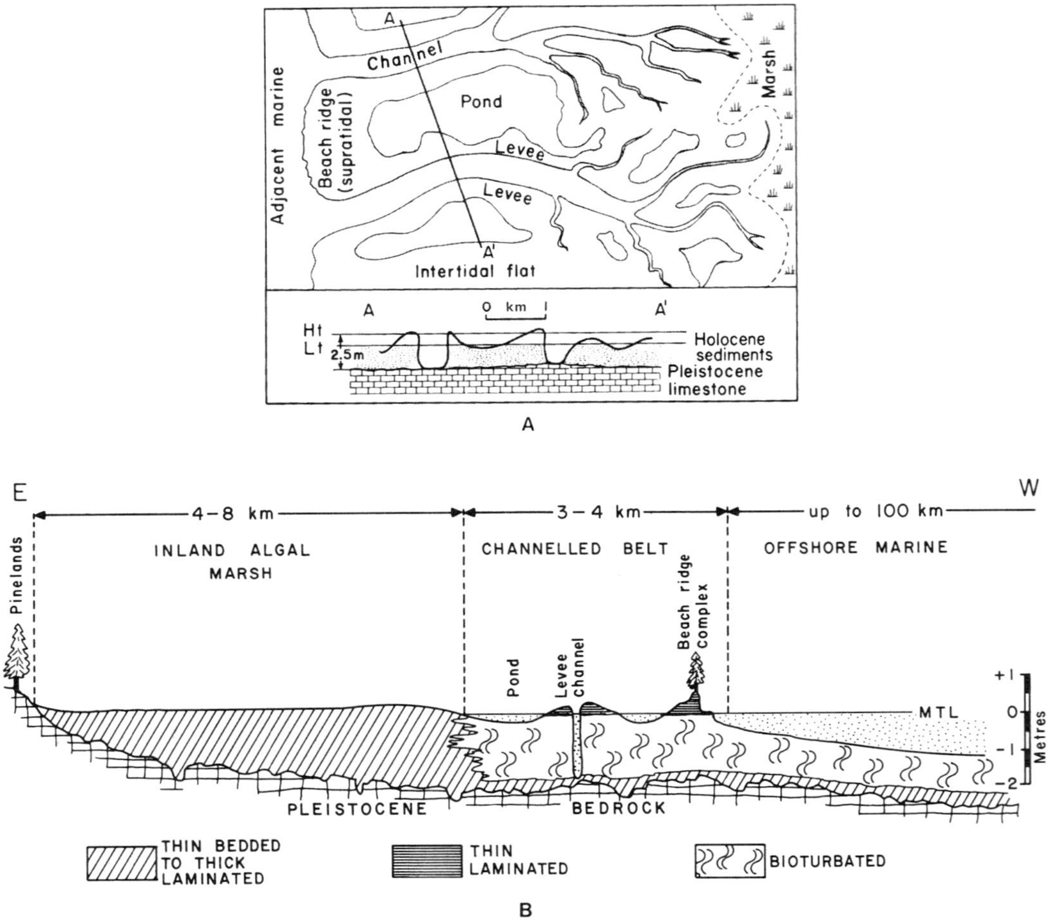

Fig. 2–29. A, Schematic map at low tide of Three Creeks area west of Andros, Bahamas (top), with idealised profile through channelled belt approximately parallel to the coast (bottom). B, Schematic cross section through the Three Creeks tidal flats. Note the laminated "cap" over bioturbated sediments in the channelled belt and landwards the thin-bedded marsh sediments. Illustration after Hardie (1977). By permission of The John Hopkins University Press.

FR vey; slikke; slikke sableuse
RE CTK; CAK

198. CARBONATE COASTAL PLAIN: the ultimate product of regression of the sea or uplift of a tidal flat. Scars of channels and some tidal ponds or lakes may characterise carbonate coastal plains.

NE karbonaat kustvlakte
DE Karbonatküstenebene

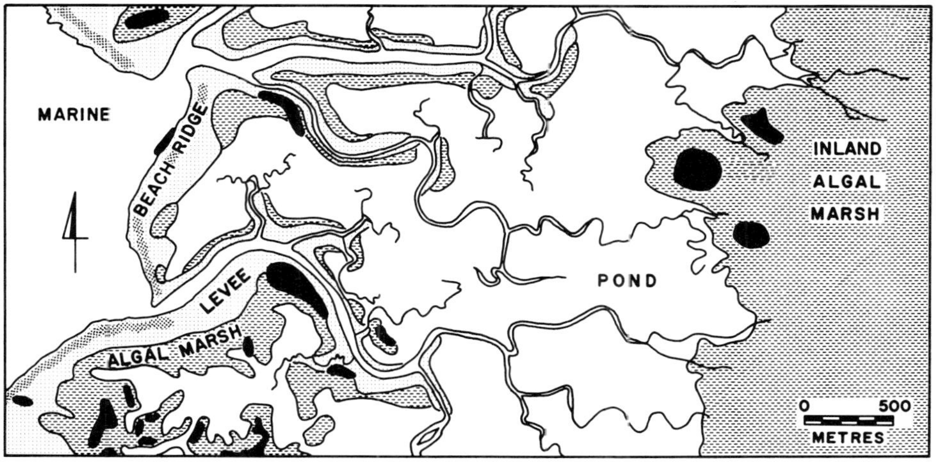

Fig. 2–30. Map showing the complex of settings at Three Creeks. Illustration after Hardie (1977). By permission of The John Hopkins University Press.

ES llanura litoral carbonatada
FR plaine littorale carbonatée; plaine côtière carbonatée
RE CTK

199. TIDAL POND OR LAKE: irregularly shaped lake in the channeled zone of a tidal flat province, immediately adjacent to the channels and their accompanying levees (see Fig. 2–31).

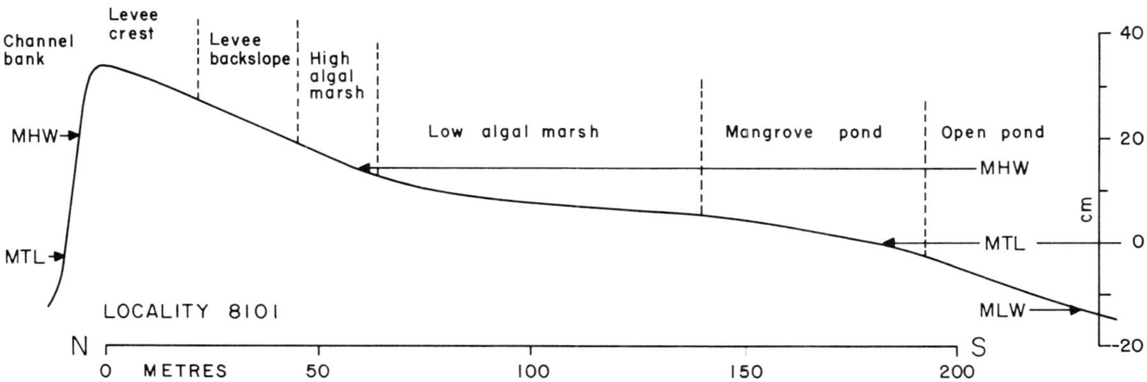

Fig. 2–31. Profile across a levee from channel to pond showing that subenvironments and settings depend on horizontal and vertical distances. Illustration after Hardie (1977). By permission of The John Hopkins University Press.

NE getijdenmeer
DE Gezeitentümpel
ES pantano, lago de marea
FR dépression en milieu intertidal
RE CTL

200. ALGAL MARSH: a marsh occurring mostly landwards in tidal flat zones and sometimes composed of "meadows" of blue-green algae. It might also form fringing algal mats immediately down the backslope of tidal channel levees, transitional to tidal ponds (see Fig. 2–32).

NE algenmoeras
DE Algenmarsch
ES marisma de algas
FR marais maritime
RE CTK

201. ALGAL MAT: an organic film of blue-green algal filamentous tissue covering sediment (see Fig. 2–33).

Note: See Horodyski and Bloeser (1977).

NE algenmat
DE Algenmatte

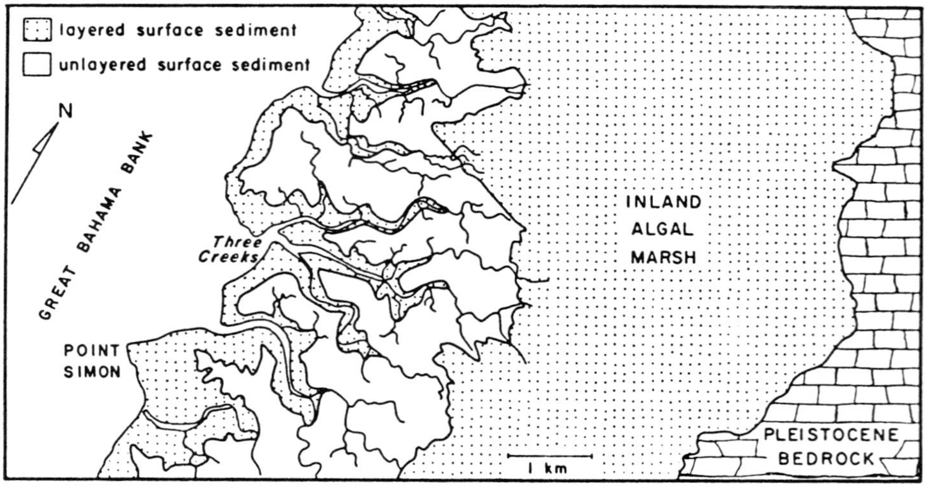

Fig. 2–32. Map of Three Creeks area showing the distribution of algal marshes and of layered and unlayered surface sediments. Illustration after Hardie (1977). By permission of The John Hopkins University Press.

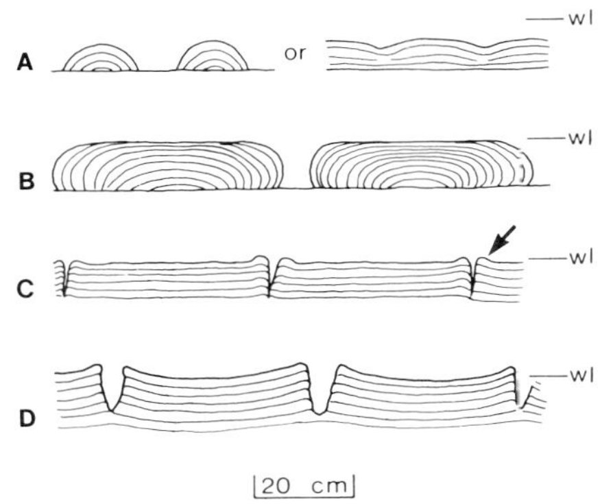

Fig. 2–33. Cross sections of *Microcroleus* algal mat (wl denotes water level). A, Mound-shaped mats grow up to the water surface and laterally merge. B, Flat-topped mounds result. C, These develop into horizontal mats that are broken by desiccation cracks. Water in these cracks encourage ongoing growth of algae, and growth ridges result (arrow). D, Greater degree of desiccation results in saucer-shaped polygons.

ES malla de algas
FR tapis algaire

202. SABKHA (SEBKHA): a supratidal flat in which evaporatic processes take place bordering tropical seas or lakes in arid climate zones (see Fig. 2–34).

Note: The term *sabkha* is also used for inland salt plains.

NE sebka
DE Sebkha
ES sebkha
FR playa; sebkha
RE CT

Shallow Marine Carbonate Settings and Bodies

Amongst marine carbonates, the "reef" *sensu lato** forms a prominent body. Carbonates associated with reefs may form other characteristic sediment masses or sediment bodies which could occur either in the deeper marine zone (below the wave base or the base of light penetration) or in the backreef setting. The transport of sediments forming the latter type of bodies could have been mechanical.

Reefs are often associated with lagoons in which protected and/or restricted carbonate masses accumulate. In reefal provinces, a forereef, a reef and a backreef zone can be distinguished.

* The meaning of reef-related terms varies considerably in various geological studies, and here a selection has been made in which the guiding principle is industrial practicality (see also "Ancient Reef Morphology").

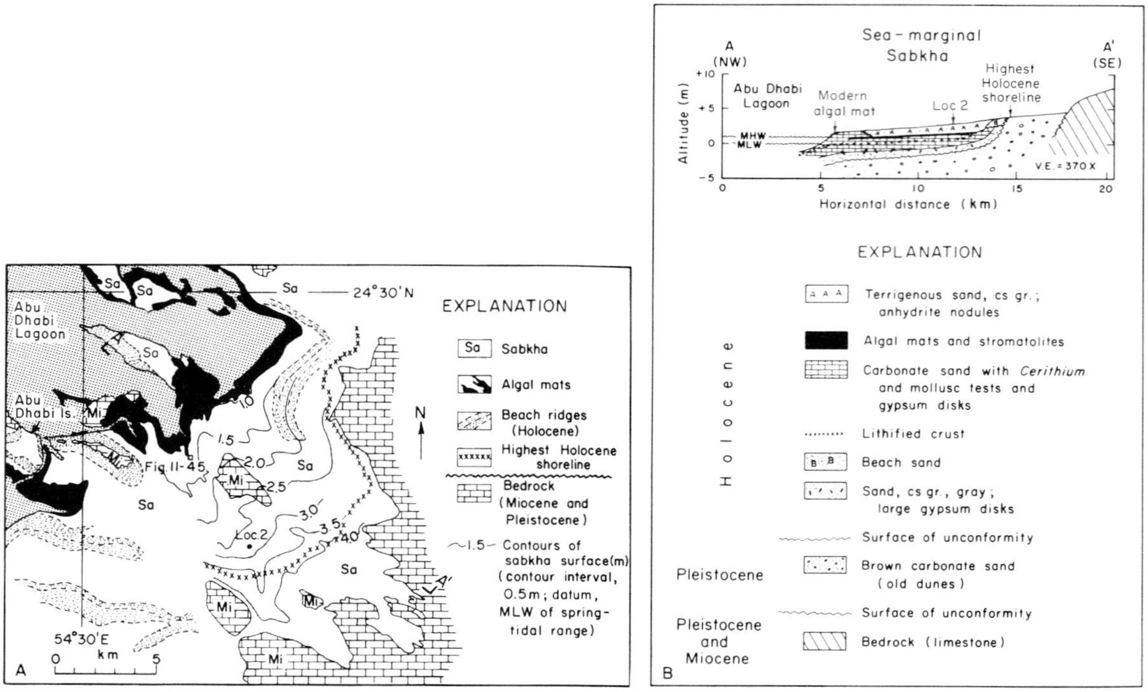

Fig. 2–34. Sabkha east of Abu Dhabi. A, Detailed map. B, Profile and section. By permission of J. Wiley & Son Inc.

Reef Classification

Darwin (1874) proposed a scheme for the origin of oceanic reefs. He recognised a trinity of reefs (fringing reef, barrier reef and atolls) in an overall deep marine environment and explained this observation in a genetic way by deducing a sequence of these sediment bodies around a subsiding central island (Fig. 2–35, top). There has been no need to replace Darwin's explanation of the geomorphology of oceanic reefs because of its essential accuracy. To Darwin's concepts Daly (1945) added the *glacial control* theory. Daly's addition is significant because it draws attention to the undeniable influence exercised by glacier-controlled water stands in reef evolution. Shelf reef morphologies, however, cannot be explained by the Darwin–Daly concepts. Such reefs, of which the Australian Great Barrier Reefs are an example, form in shallow marine seas marginal to large landmasses. Maxwell (1968) described the Australian forms in great detail. His descriptive system (Fig. 2–35, bottom) has been designed with a genetic implication that parallels the classic Darwinian sequence of fringing, barrier and atolls reefs. Maxwell's classification is based not on submergence of the reef foundation but rather on the life cycle (maturation) of the reef. More recently (1982) Hopley reevaluated the Australian Great Barrier Reef. He placed more emphasis on the antecedent karst relief, which triggers reef growth. The growth form of reefs is affected by winds, currents, tides, nutrient content of the water, light penetration and biological competition.

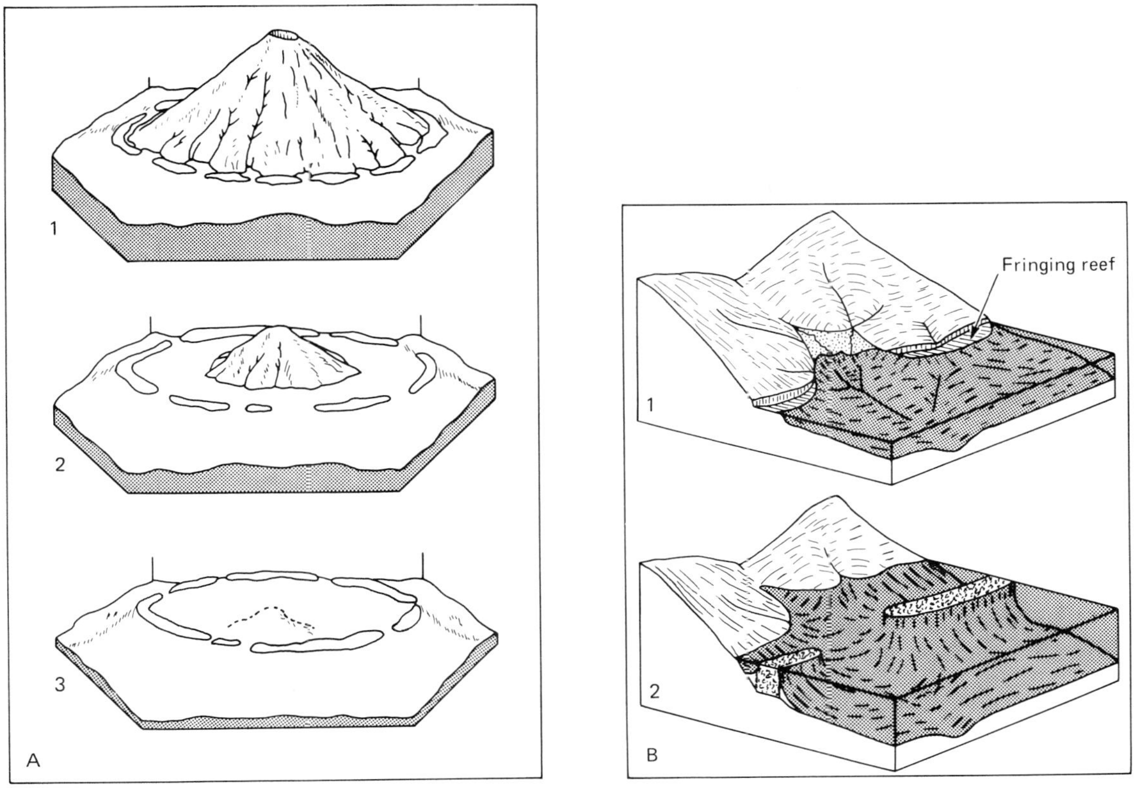

Fig. 2–35. Darwin's theory of stages in the development of atolls and barrier reefs. A, Atolls formed by the submergence of a conical volcanic island: (1) fringing reef; (2) barrier reef; (3) atoll. B, Linear barrier reef formed by the submergence of a landmass: (1) fringing reefs; (2) barrier reef and lagoon. By permission of J. Wiley & Son Inc. C, Classification of oceanic and shelf reefs. Illustration after Maxwell (1968) with permission of Elsevier Publishing Company.

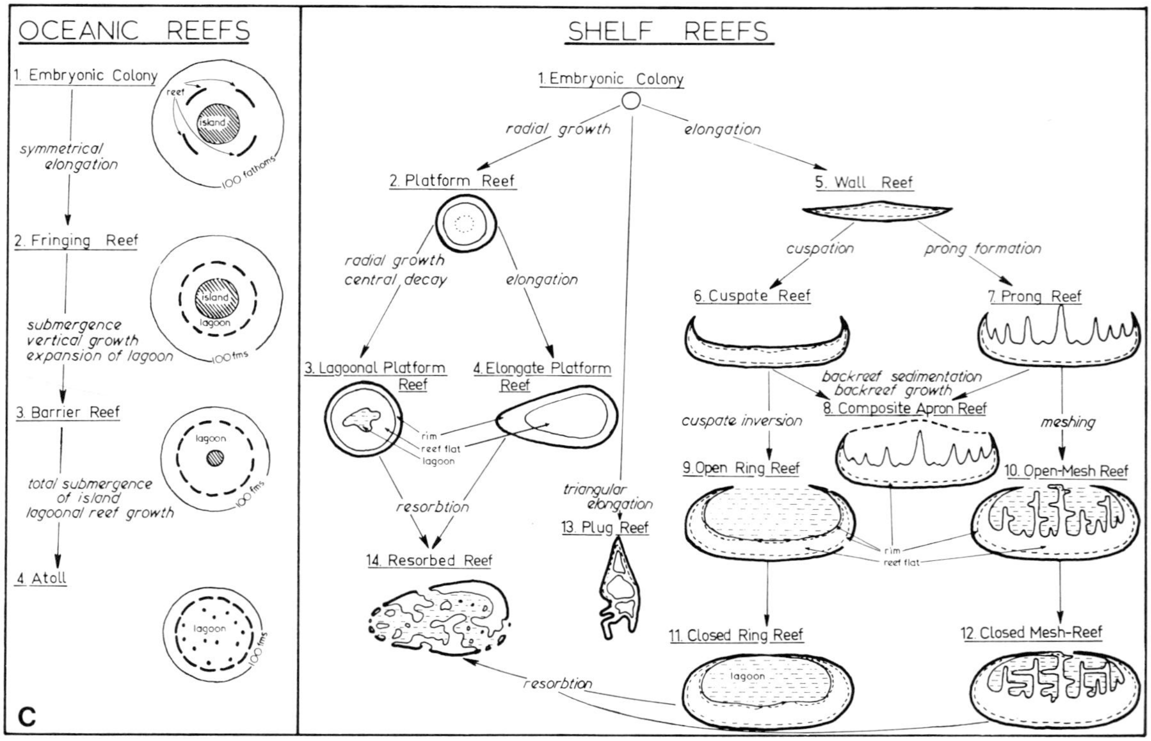

Fig. 2–35. (*Continued*)

Time, in combination with variation in the depth of the environment of deposition, is reflected in morphological maturation, as well as in growth patterns. The growth behaviour of shelf reefs reflects depth variations through time. With respect to the coastline, reefs might grow inwards or outwards. Such growth variation influences the internal distribution of facies that reflect particular reef zones. Recognition of growth behaviour (here termed *sedimentary mechanisms*) is of great importance and facilitates genetic classification of carbonate sequences in terms of shoaling/deepening or of transgression/regression.

Definition 203 (Recent Oceanic Reef)

203. ATOLL (REEF RING; RING REEF; LAGOON ISLAND): a ring-shaped coral reef appearing as a low, roughly circular coral island, or as a ring of closely spaced coral islets, encircling a shallow lagoon in which there is no pre-existing land or islands of non-coral origin and surrounded by deep water of the open sea (see Fig. 2–36).

NE atol
DE Atoll; Ringriff; Insellagune

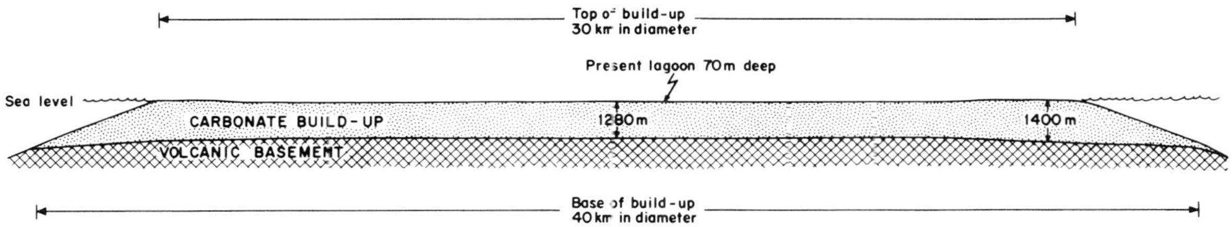

Fig. 2–36. Eniwetok Atoll, Marshall Islands (Eocene–Recent). Rate of growth since Pleistocene is 1 m per 500 years (12 m in past 4000 years).

ES atolon
FR atoll
RE OR

Definitions 204–213 (Recent Shelf Reef)

204. REEF; ORGANIC REEF (biological): a product of the actively building and sediment-binding biotic constituents which, because of their potential wave resistance, have the capacity to erect, in combination with associated detrital and bioeroded sediments, rigid, wave-resistant topographic structures (see Figs. 2–37 to 2–40).

NE rif
DE Riff
ES arrecife
FR récif
RE MSR

205. REEFAL: pertaining to a reef and its integral parts, especially to the carbonate deposits in and adjacent to a reef.

NE van een rif
DE zum Riff gehörend
ES arrecifal
FR récifal
RE MSR

206. REEFOID: resembling a reef.

NE rifachtig
DE riffähnlich
ES de tipo arrecifal
FR récifal
RE MSR

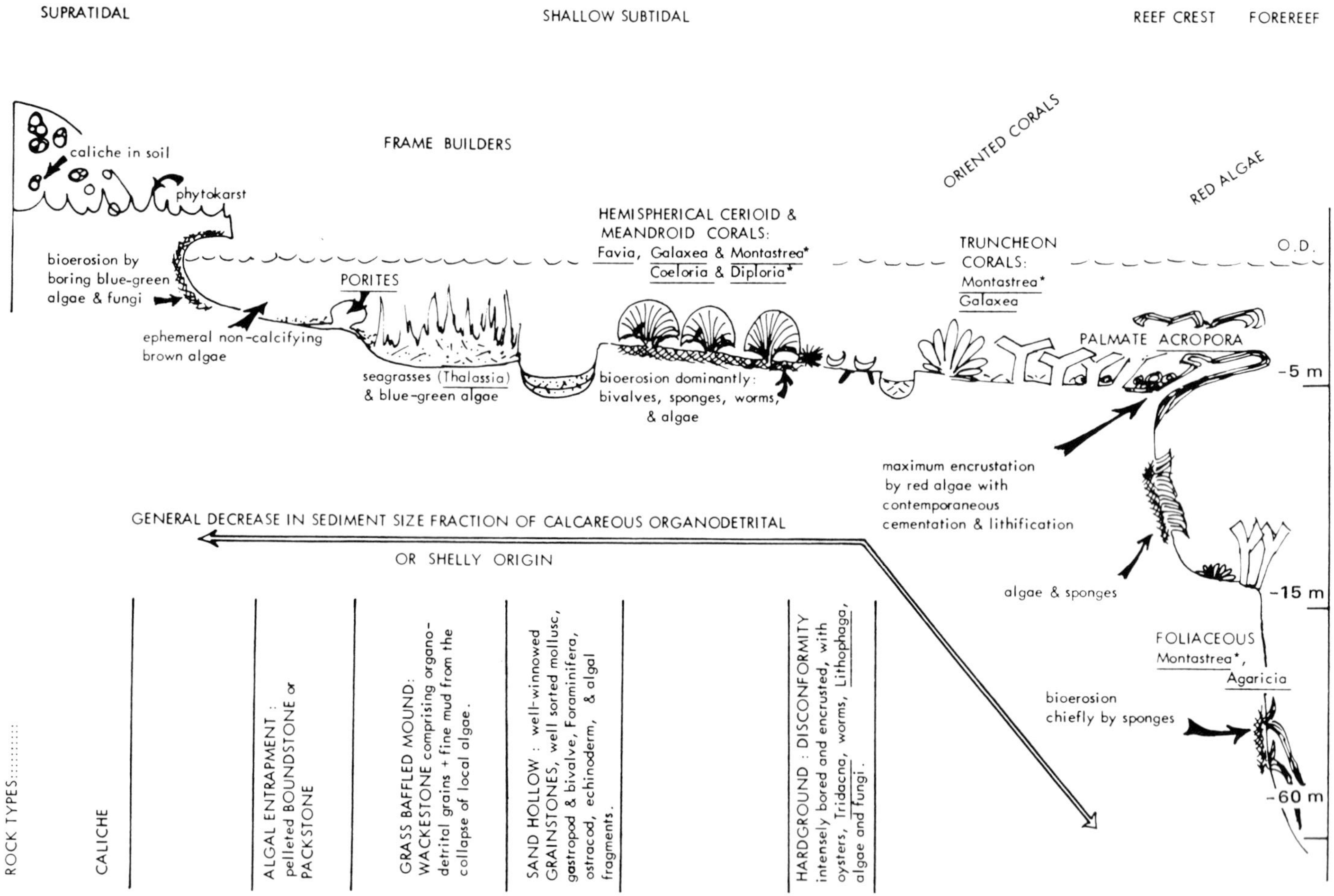

Fig. 2–37. Schematic section through an idealised reef showing relationship between frame-building organisms and sediment. Asterisks indicate taxa limited exclusively to the Atlantic. Illustration by permission of Dr. J.A.E.B. Hubbard and Elsevier Publishing Company.

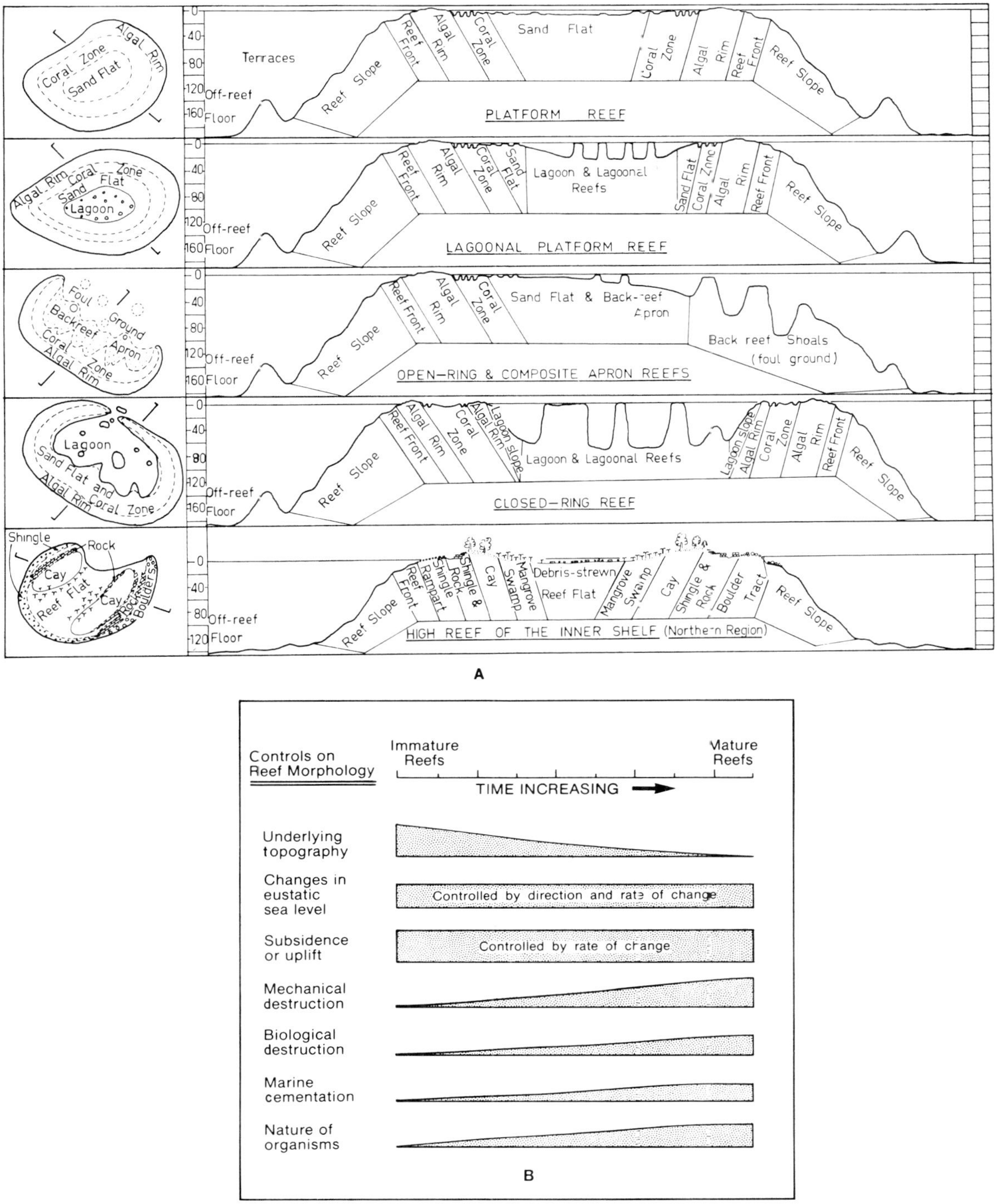

Fig. 2–38. A, Reef profiles of common reef types with characteristic physiographic zonation. Illustration after Maxwell (1968) with permission of Elsevier Publishing Company. B, Relative importance of the various processes affecting reef shape (in cross section) as a function of time. After Longman (1981). By permission of SEPM.

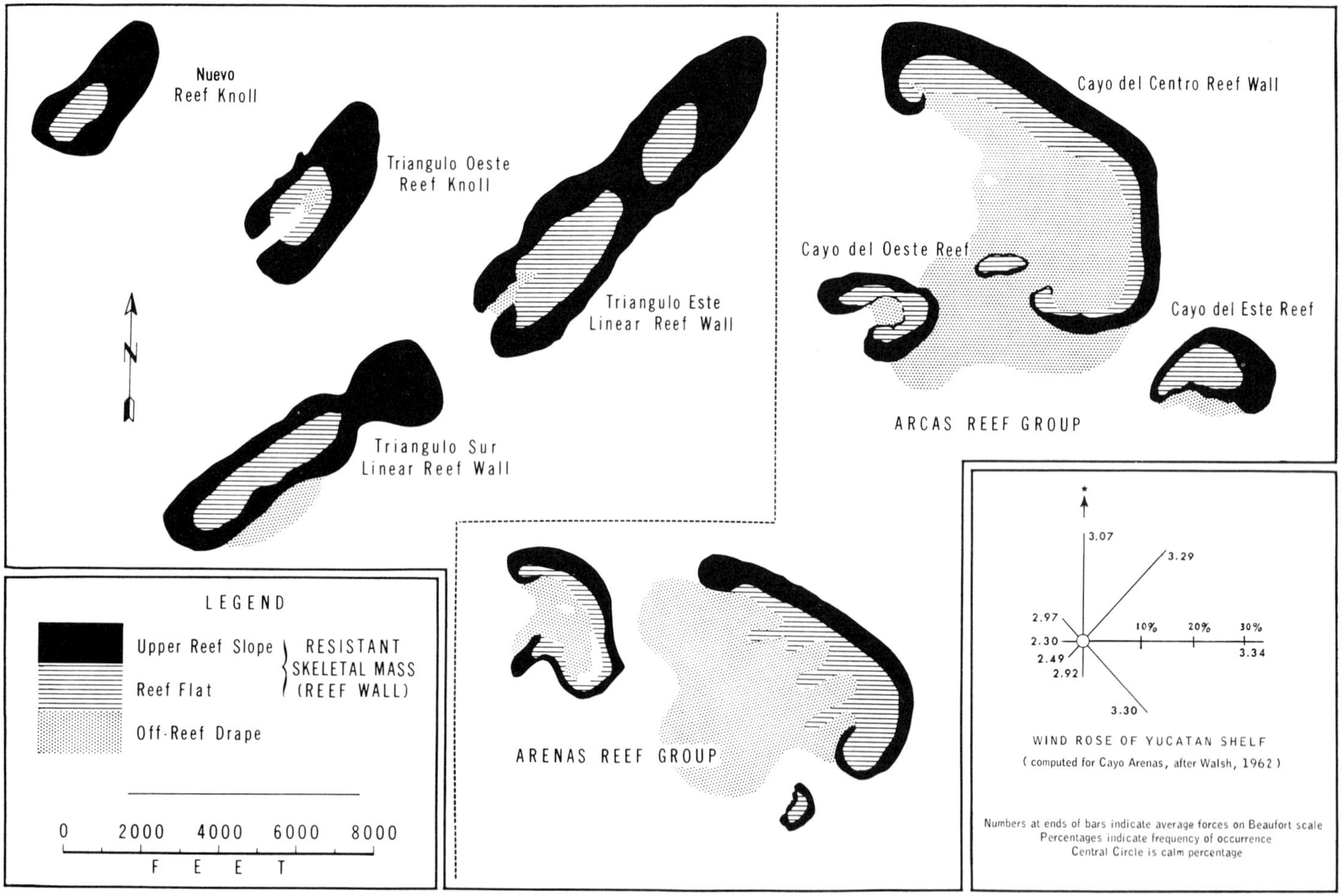

Fig. 2–39. Silhouettes of smaller reef wall structures showing relative size and orientation with respect to regional wind rosette, Yucatán shelf. From Logan *et al.* (1969). Illustration by permission of AAPG.

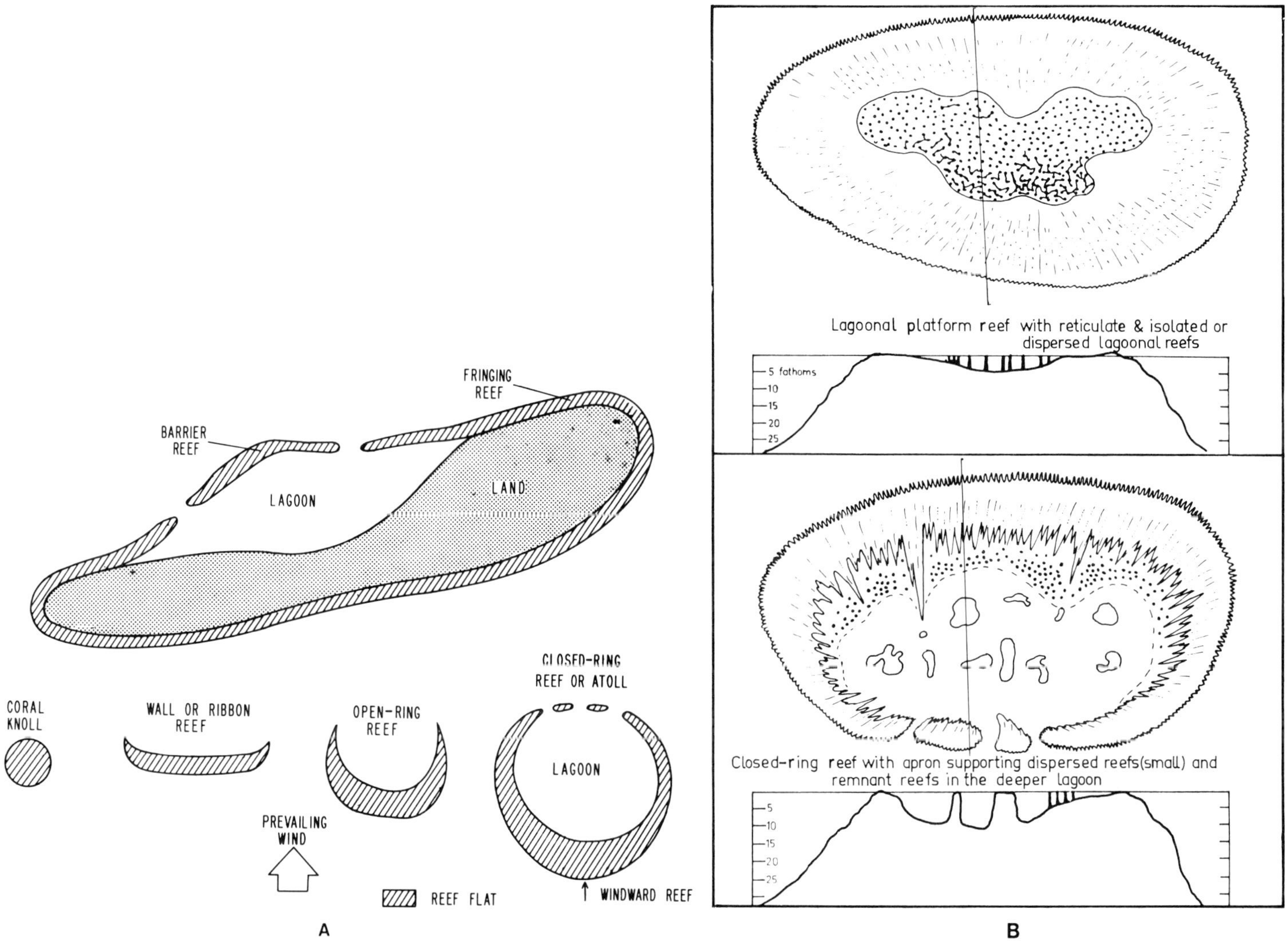

Fig. 2–40. A, Diagrammatic representation of barrier, fringing and atoll-like reefs. B, Lagoonal platform reef with dispersed lagoonal reefs and closed-ring reef with remnant reefs in the deeper lagoon. After Maxwell (1968) with permission of Elsevier Publishing Company.

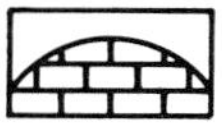 P

207. REEF PATCH (PATCH REEF): (1) a growth of coral varying in extent from several kilometres across to a mushroom-shaped growth consisting of a single large colony (see Fig. 2–41). Such reefs are formed in shelf seas usually of less than 70 m depth; they also occur in lagoons of barrier reefs or atolls (see Fig. 2–53). (2) isolated, more or less circular area of organic frame-constructed build-ups. In Recent seas, patch reefs are mainly on shelves and rise through the wave base to close to the sea surface.

Note: Lophophyllum reefs on the Blake plateau (deeper than 70 m) have also been termed patch reefs.

NE vlekkenrif
DE Fleckenriff; Kuppenriff
ES núcleo arrecifal
FR paté de corail
RE MSRP

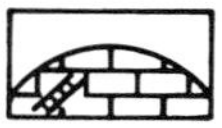

208. ALGAL REEF: an organic reef composed largely of algal remains and in which algae are the principal organisms secreting calcium carbonate (see Fig. 2–42).

NE algenrif
DE Algenriff
ES arrecife de algas
FR récif algaire
RE MSRA

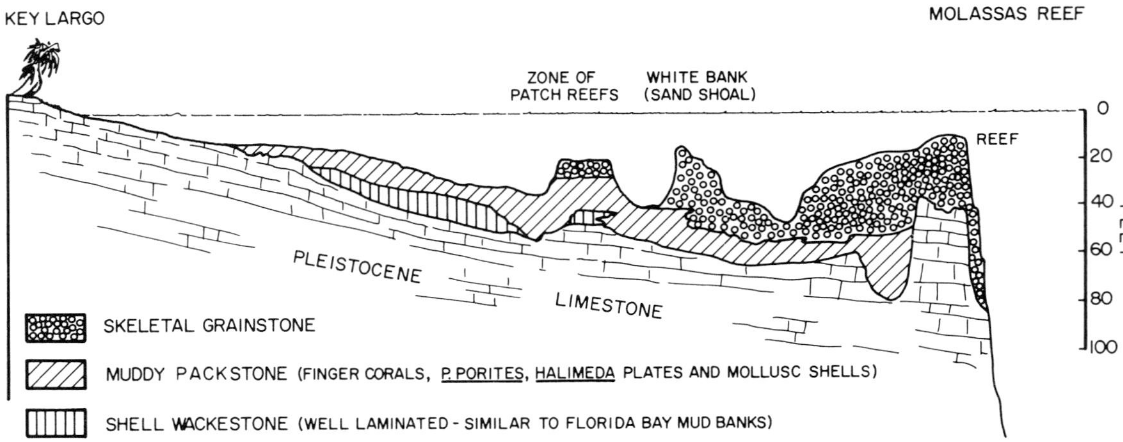

Fig. 2–41. Cross section through the Florida reef tract from Key Largo to Molasse Reef, showing the distribution of skeletal grainstones and muddy carbonates. Illustration after P. Enos with permission.

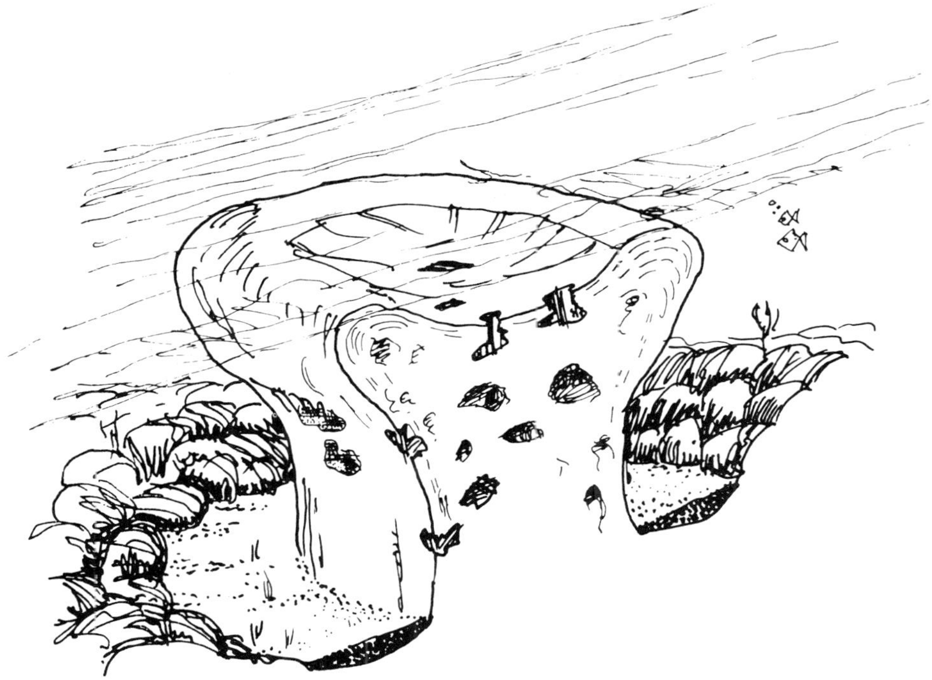

Fig. 2–42. Idealised sketch of algal cup reef, Bermuda. Illustration from Ginsburg *et al.* (1971), by permission of The John Hopkins Press.

209. HERMATYPIC CORAL: coral that lives in symbiosis with zooxanthellae and is able to construct reefs within the euphotic zone.

NE hermatypisch koraal
DE hermatypische Korallen
ES corales hermatípicos
FR coraux hermatypiques

210. AHERMATYPIC CORAL: coral that does not (or did not) build reefs. Recent examples generally lack zooxanthellae and are found at all depths, from shallow to very deep water.

NE ahermatypisch koraal
DE ahermatypische Korallen
ES corales ahermatípicos
FR coraux ahermatypiques

211. REEF CAP (CORAL CAP): a deposit of fossil-reef material overlying or covering an island or mountain.

NE rif kap
DE Riffkappe; Korallenkappe
ES techo arrecifal
FR chapeau récifal

B

212. BANK (ORGANIC BANK): a carbonate build-up consisting of skeletal elements formed in place or of detrital organic sediment or mud accumulated in place by trapping or baffling, but also in part through mechanical piling by waves and currents (see Figs. 2–43 and 2–44).

NE bank
DE Bank
ES banco
FR banc
RE MSP

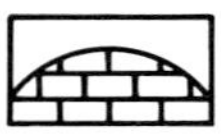
B,P

213. SEDIMENT PILE: an accumulation of sediment in almost any shape whose internal composition shows it to originate mainly through mechanical piling of transported particles (e.g., dunes, bars, spits, tidal deltas). The term *bank* has also been used to mean a sediment pile.

NE sediment hoop
DE Bank
ES banco
FR banc
RE MSD

Definitions 214–215 (Recent Reef Flat)

214. REEF FLAT: a stony platform of coral fragments and coral sand formed on the inner or lagoonal side of a coral reef and generally exposed at low tide (see Fig. 2–45).

NE droogvallend koraalplatform
DE Riffplatte; Riff-Fläche
ES llanura arrecifal
FR plateau récifal; plateforme récifale
RE MSRF

215. ALGAL RIDGE: a low ridge or elevated margin at the seaward (outer) edge of a reef flat, composed of calcium carbonate secretions of actively growing calcareous red algae (see Fig. 2–46). It may occur independently of reef flats and form algal reefoid structures.

Note: This is specifically a feature of present-day coral–algal reefs.

NE algendijk
DE Algenbarre; Algenrücken
ES cresta algal
FR trottoir à algues

Definitions 216–218 (Recent Upper Reef Slope)

216. REEF FRONT: the upper part of the seaward slope of the reef, between the dwindle point of abundant living coral and coralline algae and the reef edge (see Fig. 2–47).

NE riffront
DE Riff-Front
ES frente arrecifal
FR front récifal

217. REEF EDGE: the seaward margin of the reef flat, commonly marked by surge channels.

NE rifrand
DE Riffrand; Riffkante
ES margen arrecifal
FR rebord récifal

218. REEF-FRONT TERRACE: a shelf or benchlike surface which is sometimes the result of erosion. Such terrasses may be covered with organic growth. They occur below and in front of the reef edge, slope seawards (to a depth of 15 to 30 m) and may reflect ancient sea level stands.

NE riffront terras
DE Riffront-Terrasse
ES terraza arrecifal
FR terrasse récifale

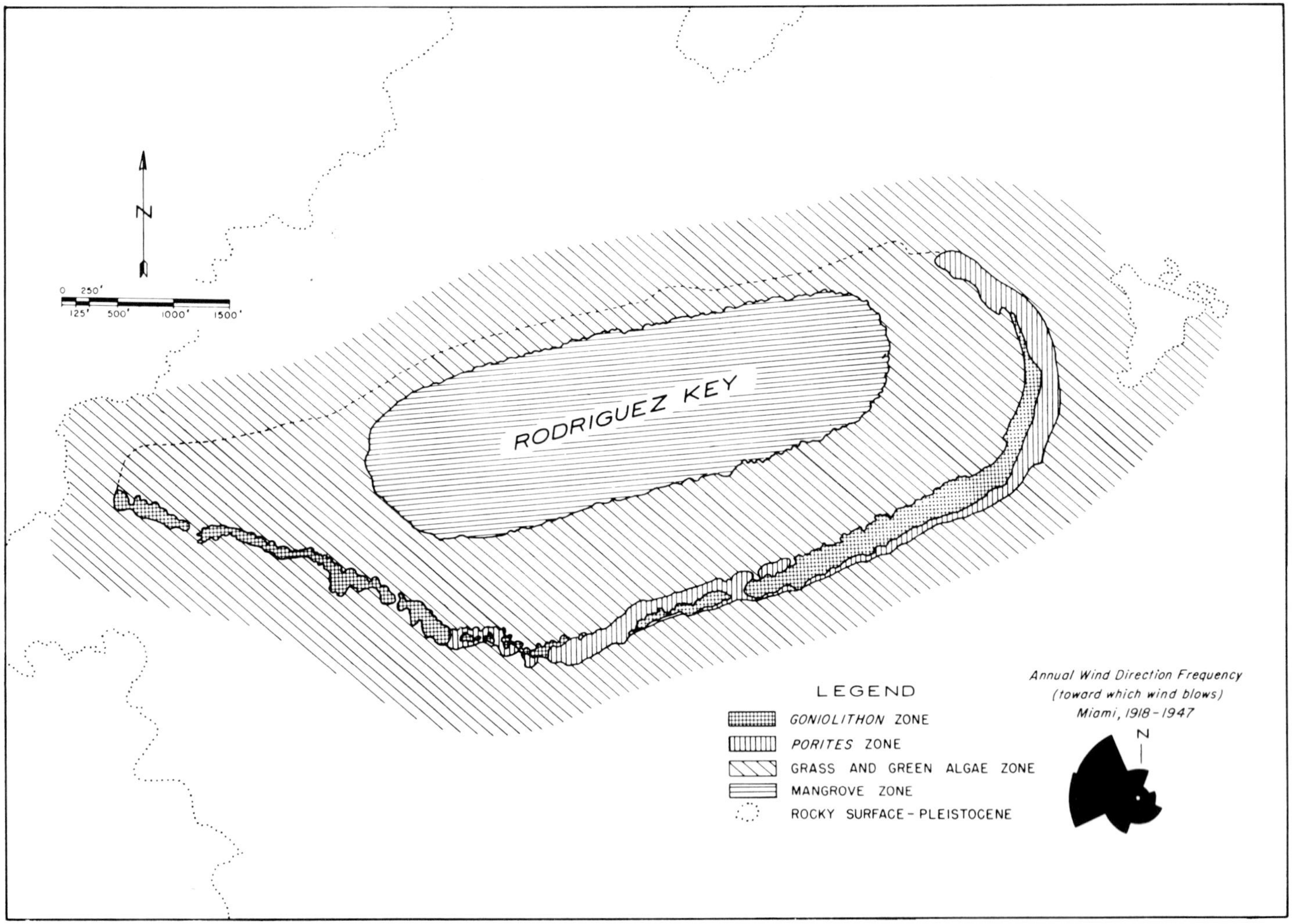

Fig. 2–43. Rodriguez Bank, east of the Florida Keys, demonstrates the development of a narrow reefoid rim (a windward bank margin) composed of zones with small branching corals (e.g., *Porites*) and branching coralline algae (e.g., *Gonialithon*). Illustration after Turmel and Swanson (1977) with permission of Soc. Econ. Palaeontol. Mineral.

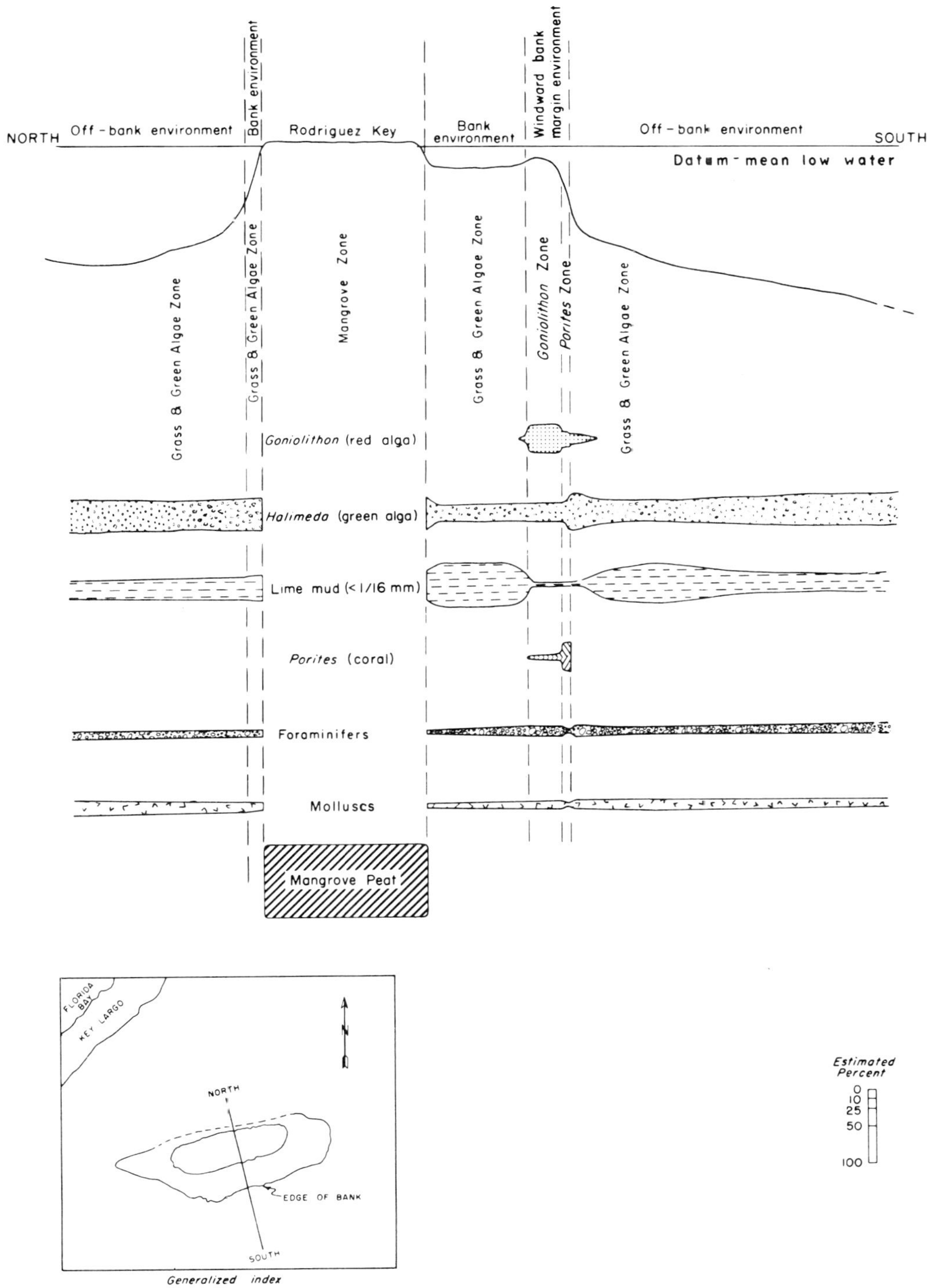

Fig. 2–43. *(Continued)*

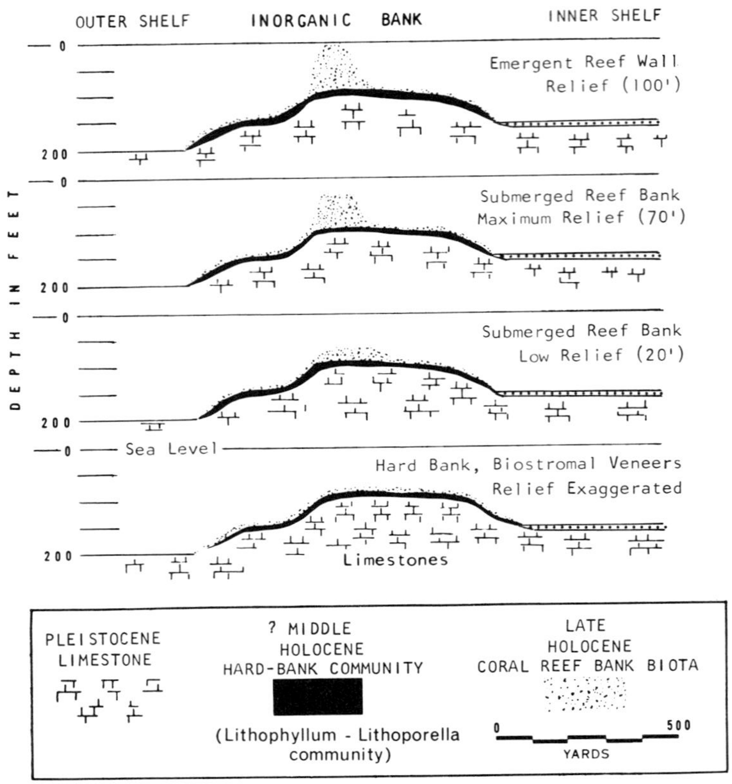

Fig. 2–44. Diagrammatic cross sections showing relation of organic structures to rocky inorganic bank nuclei. From Logan *et al.* (1969) by permission of AAPG.

GRAIN SIZE: SAND AND GRAVEL, < 1–2 % SILT AND CLAY

COMPOSITION*:	CARIBBEAN (50 coral species)	PACIFIC (700 coral species)
CORAL FRAGMENTS	15–35 %	15–36 %
CORALLINE ALGAE FRAGMENTS	10–33 %	23–52 % (90% on Siapan)
MOLLUSCS	5–22 %	12 %
HALIMEDA	16–40 %	2–10 %
BENTHONIC FORAMINIFERA	2–13 %	10–23 %

Fig. 2–45. Approximate composition of Carribean and Pacific Recent reef flat sediments. The proportions given do not necessarily reflect the relative abundances of the living organisms on the reef.

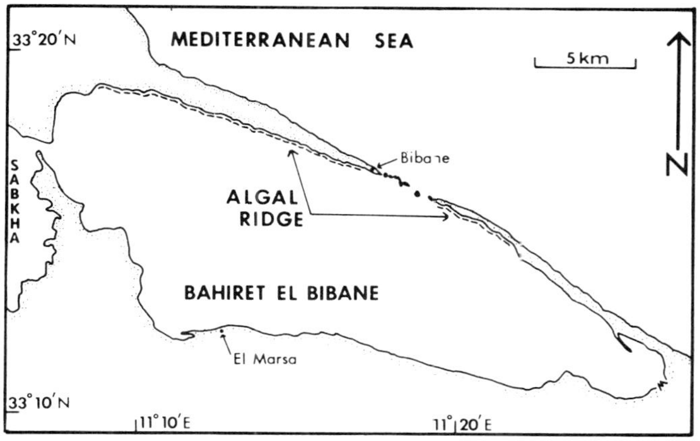

Fig. 2–46. Mapped extent of an algal ridge, Tunesia. Illustration by permission of the Soc. Econ. Palaeontol. Mineral.

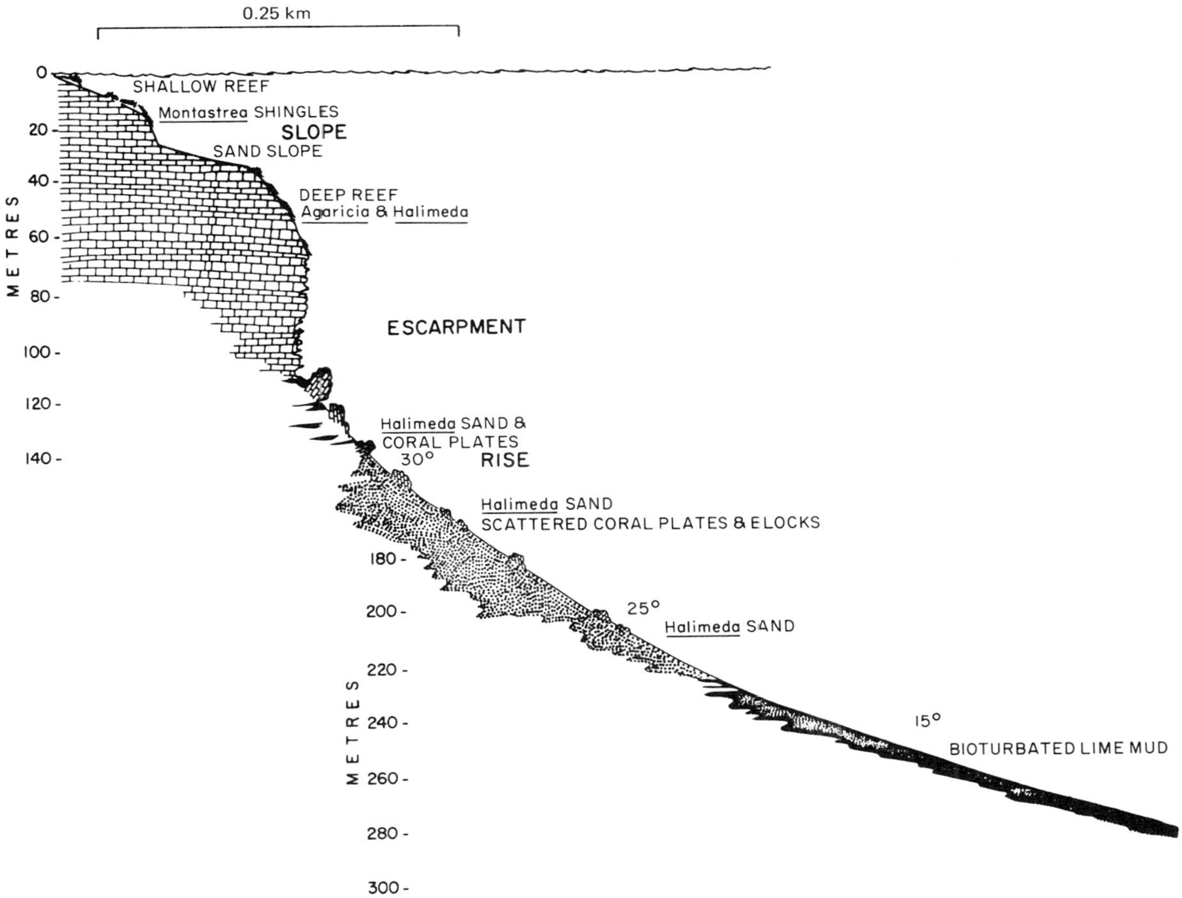

Fig. 2–47. Profile of the platform margin off the Belize barrier reef from 0 to 300 m, illustrating the position of deep reef communities. Illustration after Ginsburg and James (1974), with their kind permission.

Definitions 219–223 (Recent Lower Reef Slope)

219. REEF FLANK: the part of the reef that surrounds, interfingers with and locally overlies the reef core. It is often indicated by massive or medium beds of reef talus dipping steeply away from the reef core.

NE rifflank
DE Riff-Flanke
ES flanco arrecifal
FR bord du récif; talus récifal
RE MSRD

220. REEF SLOPE: the face of the reef, rising from the off-reef floor (see Figs. 2–48 and 2–49).

NE rifhelling
DE Riffhang
ES talud arrecifal
FR pente récifale

221. REEF WALL: a wall-like upgrowth of living reef-building organisms and the skeletal remains of dead coral and other skeletal sessile organisms reaching intertidal level, where it acts as a partial barrier between adjacent environments (see Fig. 2–50).

NE rifmuur
DE Riffgerüst; Riffmauer; Riffwall

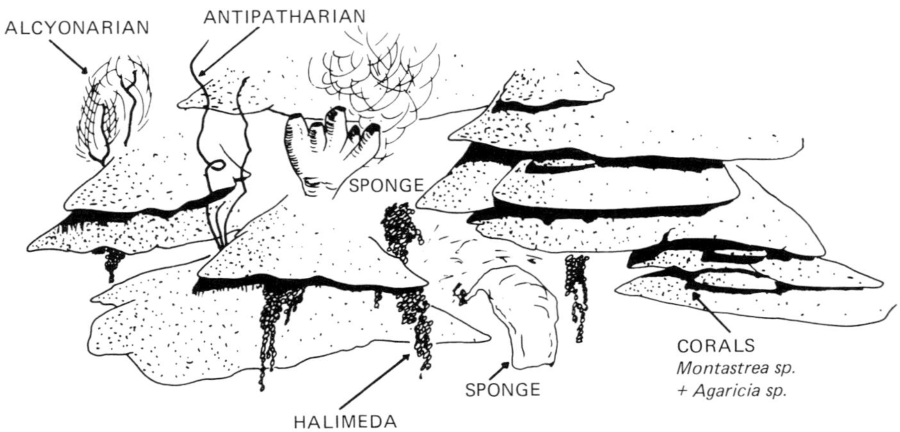

Fig. 2–48. Sketch illustrating the growth forms of the sponges, corals and algae that form the deep reef community. Illustration after Ginsburg and James (1974), with their kind permission.

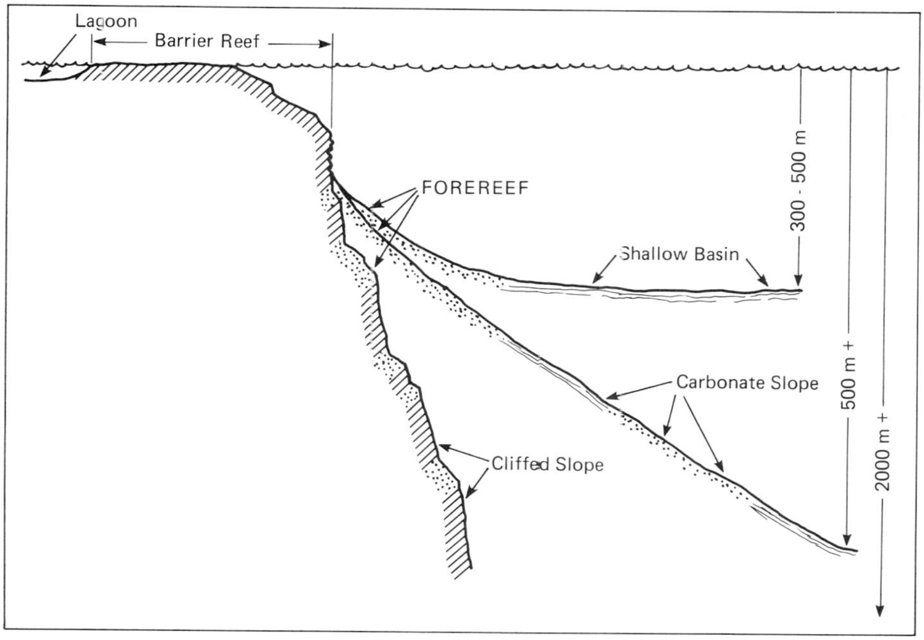

Fig. 2–49. Three different styles of deep platform margins. The forereef grades directly in a shallow basin; it may be the upper part of a larger, gently dipping slope or the upper part of a steep, dipping slope. Illustration courtesy of the International Assoc. of Sedimentologists.

ES pared arrecifal
FR paroi récifale, tombant récifal

222. REEF TALUS: displaced material covering the reef slope and derived from the reef core or wall (see Fig. 2–51).

NE riftalud
DE Riff-Halde; Riff-Schutt
ES depósitos de talud arrecifal
FR talus récifal

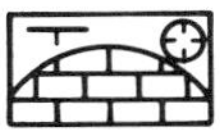
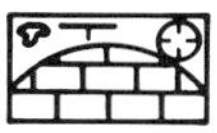

223. REEF BRECCIA: a rock formed by the consolidation of limestone fragments, broken off from a reef by the action of waves and tides, but also by biochemical and biological activity.

NE rif breccie; rifkalkbreccie
DE Riff (kalk) breccie
ES brecha arrecifal
FR brèche récifale

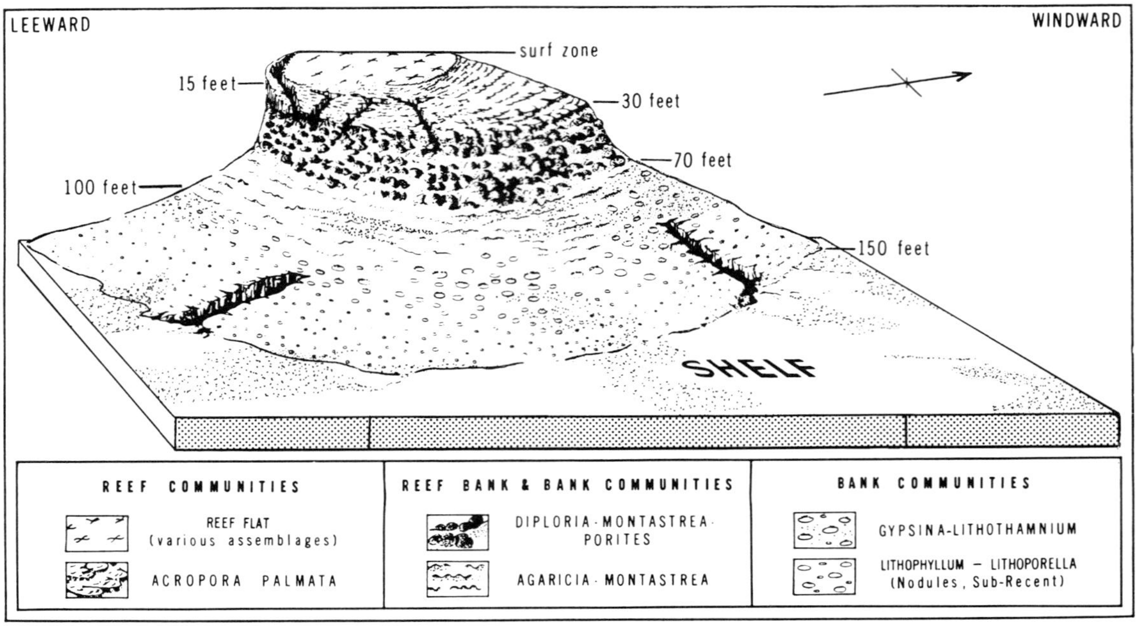

Fig. 2–50. Diagrammatic representation of reef wall in a solitary knoll structure (Neuvo Reef, Yucatán shelf), illustrating the distribution of community zones and the morphology of wall from reef flat to floor of surrounding shelf. The organic structure is based on an inorganic bank foundation, which rises to ~100-ft depth. Reef flat is horizontal surface; upper reef slope, *Acropora palmata* community zone; lower scarp, *Diploria–Montastrea–Porites* community zone; basal slope, *Agaricia–Montastrea* community zone. From Logan *et al.* (1969), by permission of AAPG.

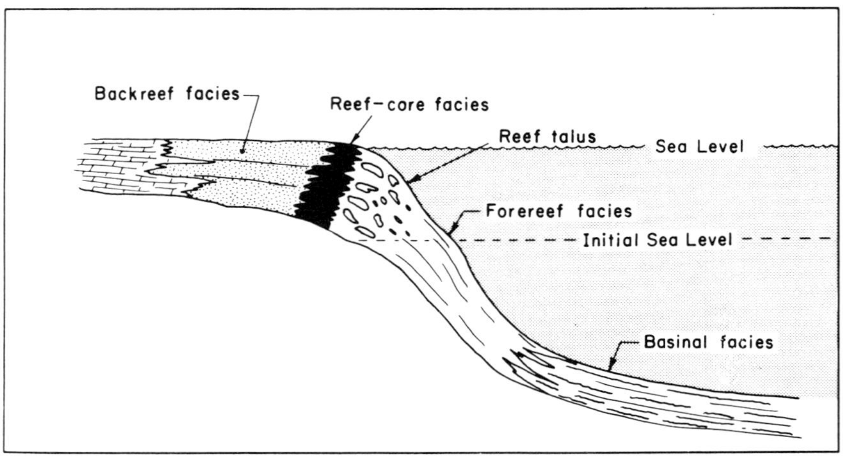

Fig. 2–51. Idealised profile and section (not to scale) across shelf-edge reef. Reef core (black), backreef (stippled), forereef facies. Illustration after Friedman and Sanders (1978) by permission of John Wiley & Sons, Inc.

Definition 224 (Recent Normal Marine Shelf)

224. REEF DETRITUS: fragmental material eroded off the organic reef and deposited nearby.

NE rifpuin
DE Riffdetritus; Riffschütt
ES material detrítico arrecifal
FR éboulis récifaux; débris récifaux

Definitions 225–227 (Miscellaneous Terms Related to Recent Forereef/Off-reef Area)

225. FOREREEF: the seaward side of a reef, commonly a steeply dipping slope with deposits of reef talus.

NE voorste rifdeel; rifhelling
DE seewärtige Riffseite; Vorriff
ES arrecife externo
FR récif externe

226. OFF-REEF: the site of deposition of some reef debris, predominantly basinal, between isolated build-ups or in front of the forereef belt.

NE buiten het rif
DE riffnahe (Sedimentation)
ES ante-arrecife
FR avant-récif

227. FARO: ringlike organic accumulation with shallow central lagoon located shelfward of a barrier reef trend (off-reef).

NE faro
DE Faro
ES faro
FR faro

Definitions 228–232 (Recent Lagoon Slope)

228. LAGOON CLIFF (LAGOON SCARP; LAGOON SLOPE): the seaward border zone or side of a lagoon, sloping downward from the lagoon margin to the lagoon floor.

NE lagune helling
DE Lagunensteilrand; Lagunensteilhang; Lagunensteilflanke

ES escarpe de laguna
FR falaise du lagon; escarpement du lagon; pente du lagon

229. ALGAL RIM (TROTTOIR; CORNICE): a low, slight rim built by actively growing calcareous algae on the lagoonal side of a leeward reef or on the windward side of a reef patch in a lagoon.

NE algenrand
DE Algenrand
ES borde algal
FR trottoir algaire; récif algaire frangeant

230. LAGOON BEACH: the sandy fringe on the inner, protected side of a reef island, facing the lagoon.

NE lagune strand
DE Lagunen-Strand
ES playa de laguna
FR plage du lagon

231. LAGOON SHELF: the part of the sea bottom adjacent to the reef at the lagoon side. It is usually covered with lime sand or lime mud but is characterised by sedimentation conspicuously exceeding organic growth.

NE lagune vlakte
DE Lagunen-Schelf
ES "lagoon-shelf"
FR plateau du lagon

232. LAGOON MARGIN: the lagoonward margin of the lagoon shelf or of the reef flat along those parts of the reef that lack islands.

NE lagune rand
DE Lagunen-Schelfrand
ES margen de laguna
FR marge du lagon

Definitions 233–239 (Recent Lagoon Floor)

233. LAGOON (LAGUNE; LAGUNA): a shallow expanse of seawater partly or completely separated from the open sea by a low, narrow, elongate strip of land such as a reef, barrier or island (see Fig. 2–52).

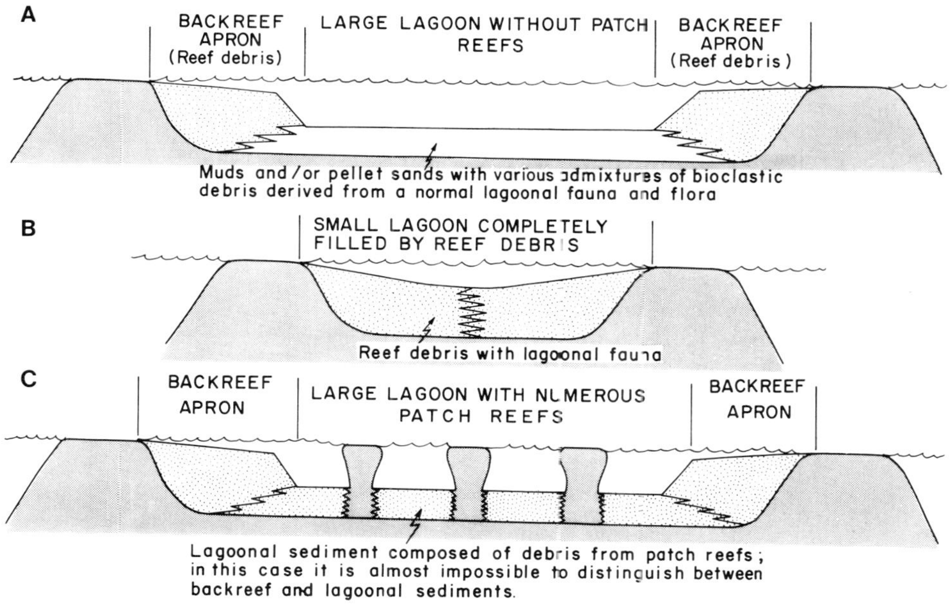

Fig. 2–52. Various ways in which lagoons can be filled up with lime mud (A) or sands (B, C). Sand is generated mainly by the reef and the interlagoonal patch reefs.

NE lagune
DE Lagune
ES laguna
FR lagon (milieu récifal); lagune (milieu non récifal)
RE ML

234. LAGOON FLOOR: the undulating to nearly level bottom of a lagoon, often encircled by the lagoon slope and/or the lagoon cliff.

NE lagune bodem
DE Lagunenboden
ES fondo de laguna
FR fond du lagon

235. LAGOON FLAT: the nearly horizontal reef flat located lagoonward of the lagoon beach.

NE lagune vlakte
DE Lagunenfläche
ES llanura de laguna
FR replat

236. BARRIER LAGOON: (1) a lagoon roughly parallel to the coast and separated from the open ocean by a strip of land or by a barrier reef; (2) a lagoon encircled by coral islands or coral reefs.

NE barrière lagune
DE Barrierelagune
ES laguna barrera
FR lagon de barrière récifale

237. LAGOON CHANNEL: (1) the expanse of deep water separating a reef from the neighbouring land; (2) a pass through a reef and into and through a lagoon.

NE lagunekanaal
DE Lagunenkanal
ES canal de laguna
FR chenal; passe

238. LAGOON ISLAND: (1) an island rising from within the lagoon of a composite atoll or a barrier reef (see Fig. 2–53); (2) atoll.

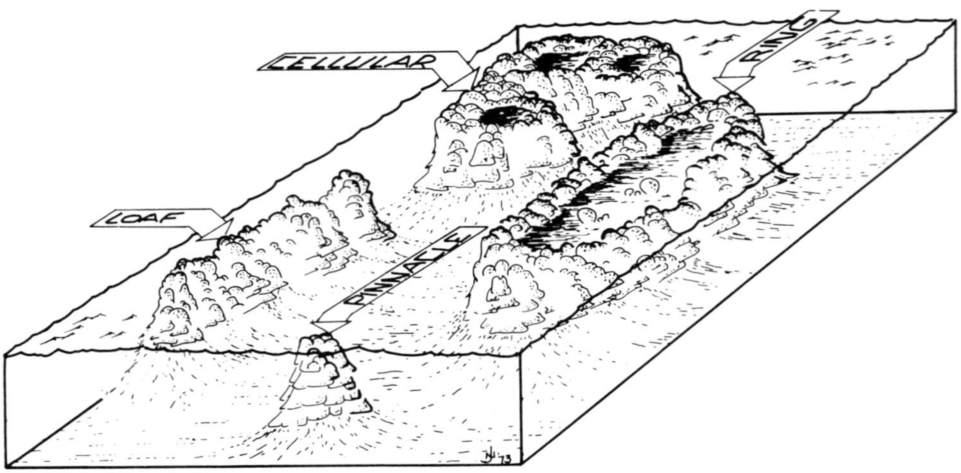

Fig. 2–53. Principal forms of lagoonal reefs. Pinnacle (knoll and patch): generally 10–100 m wide, 3–20 m high. Loaf: ~50 m wide, up to 300 m long, 4–8 m high. Cellular: irregular network of branched arrays of various dimensions; relief less than 10 m. Ring (faro, rhomboid shoal): Subcircular to polygonal, 2–8 km long and 15–30 m high. Smaller reefs are flat topped with coral growth concentrated on the margins surrounding a sand-floored center; larger rings often have a deeper central lagoon with pinnacle reefs. By Ginsburg and James (1974) with permission of the authors.

NE eiland met lagune; atol
DE Laguneninsel
ES isla de laguna; atolón
FR île du lagon

239. REEF PINNACLE (PINNACLE REEF): a small patch reef, consisting of coral growing sharply upward (slopes from 45° to nearly vertical), often within an atoll lagoon, usually rising close to the water surface (see Fig. 2–54).

Note: See Shaver (1977).

NE puntrif
DE Säulenriff; Turmriff
ES pináculo arrecifal
FR pinacle corallien; pâté de corail

Definition 240 (Miscellaneous Term Related to Backreef Area)

240. BACKREEF: the landward side of a reef, including the area and the contained deposits (partly or even mainly terrigenous) between the reef and the mainland.

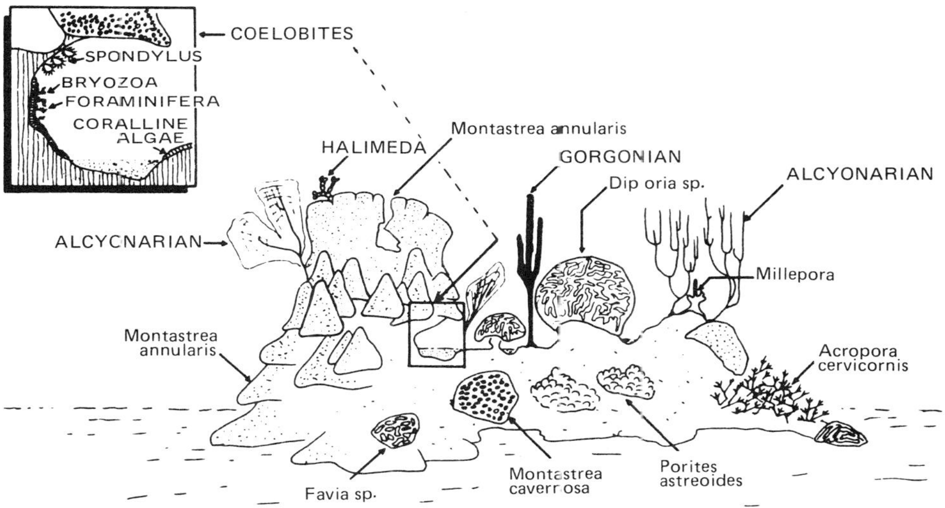

Fig. 2–54. Sketch of forms and distribution of the principal constituents of a lagoon reef community. Illustration after Ginsburg and James (1974), with their kind permission.

NE binnenrif
DE Rückriff; Hinterriff; Riffrückseite
ES arrecife interna
FR arrière récif; récif interne

Definitions 241–244 (Recent Reef Association)

241. BARRIER REEF: (1) a long, narrow coral reef roughly parallel to the shore and separated from it at some distance by a lagoon of significant depth and width (see Fig. 2–55); (2) the limestone produced by the consolidation of materials in a barrier reef.

NE barrièrerif
DE Barrierriff; Wallriff
ES arrecife de barrera
FR récif barriére; barriére récifale

242. FRINGING REEF: a curvilinear belt of sediments produced by organic accumulation and built directly out from the coast.

NE franjerif
DE Saumriff
ES arrecife costero
FR récif frangeant

243. REEF COMPLEX: a cluster of individual reefs. Refers also to the total of the various reef zones (e.g., forereef, reefflat, backreef).

NE rifcomplex
DE Riffkomplex
ES complejo arrecifal
FR complexe récifal

244. REEF TRACT: a general term indicating the zone in which reef-building organisms occur in the offshore shelf area (see Fig. 2–56).

NE rifzone
DE Riffbereich; Riff-Zone
ES zona arrecifal
FR zone récifale

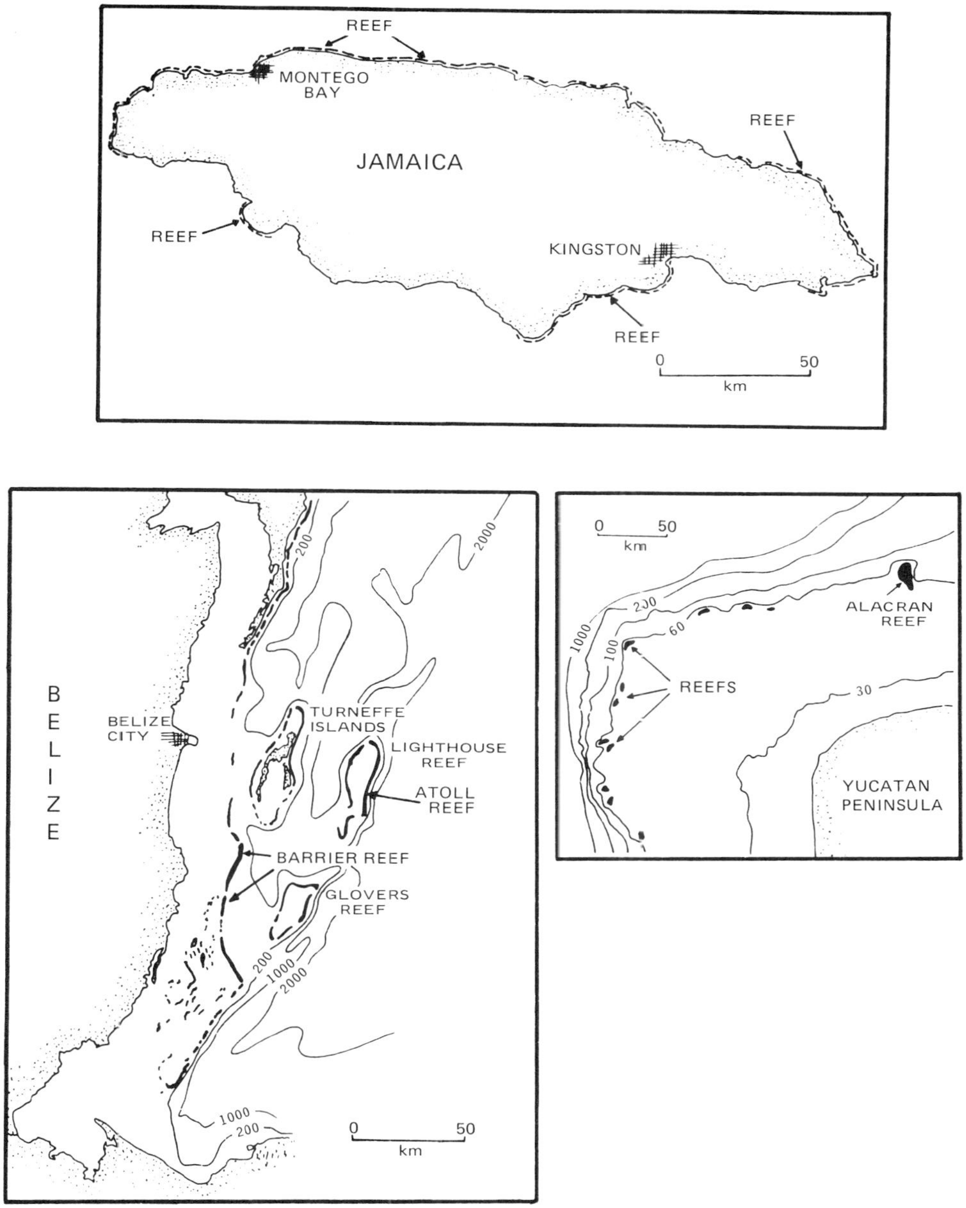

Fig. 2–55. Three types of zoned marginal reefs: Jamaica, fringing reef (after Goreau, 1959); Belize, barrier and atoll reefs (after Wantland and Pusey, 1971); Yucatán, isolated build-up capped by zoned reefs at the shelf margin (after Logan *et al.*, 1969). Illustration after Ginsburg and James (1974), with their kind permission.

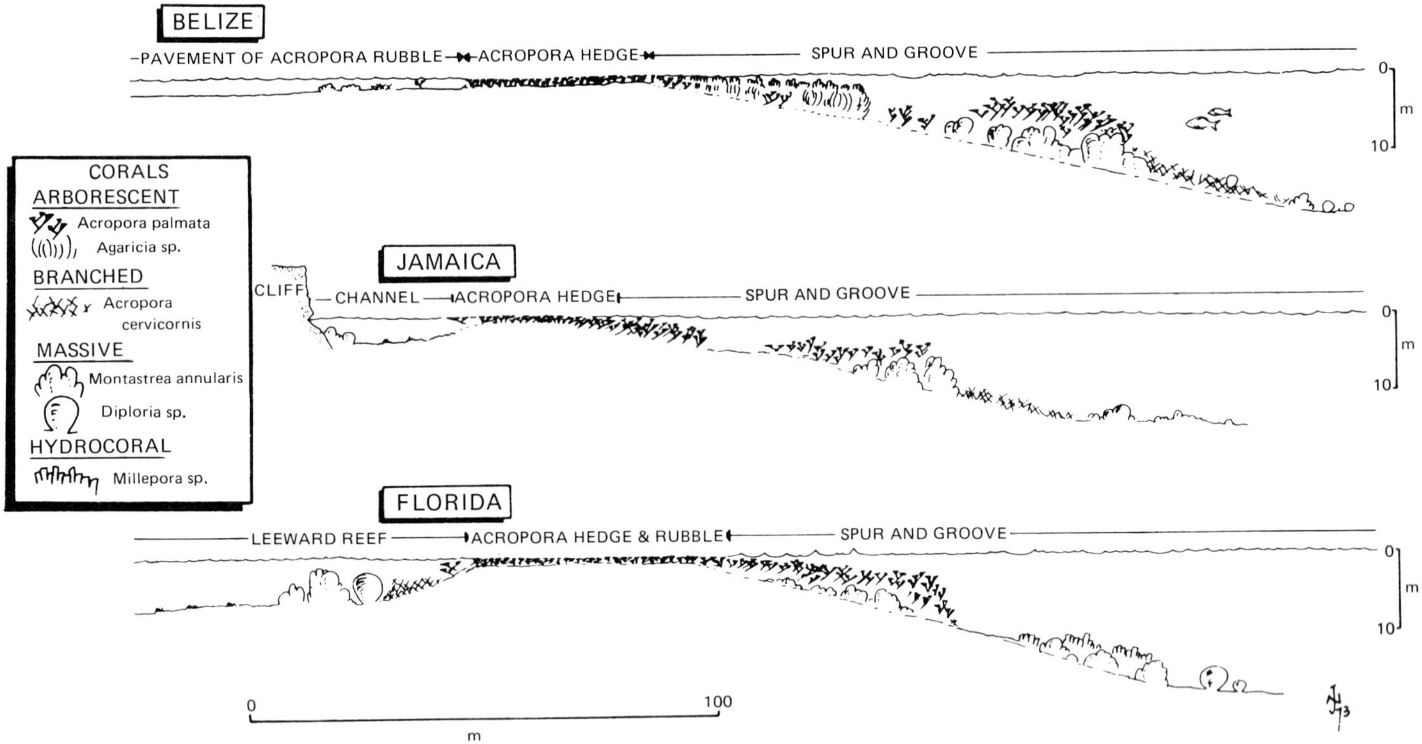

Fig. 2–56. Profiles showing the morphology and zonation across three zoned marginal reefs (see also Definition 241). Illustration after Ginsburg and James (1974), with their kind permission.

2–5 CARBONATE DEPOSITIONAL SETTINGS AND BODIES (ANCIENT)

Introduction

An increase in our understanding of the formative processes of carbonate build-ups, partly as a result of studies of Recent reefs, has resulted in a redundancy of the terminology describing such sedimentary bodies. The terms in this section are a recommended selection from a multitude of terms. The following considerations guided the compilation:

1. Expressions describing the shape of carbonate bodies are separated from those describing their internal composition.
2. Discreet small-scale limestone bodies and large regional carbonate masses may be referred to with the same name. Modifiers are then used to distinguish the two types.
3. The term *reef* has evolved rapidly and its meaning has been considerably modified and restricted. Consequently the term should always be used with a modifier (see Heckel, 1974).

Wilson (1975) discussed the occasionally cumbersome but worthwhile distinction between descriptive and partly or entirely genetic terms of reefs, and his remarks are applied here.

After the external shapes and compositional variations adopted by ancient build-ups are described, the growth behaviour of such carbonate masses is delineated, as reflected in their internal facies distribution. Sedimentary tendencies are deduced from standard vertical sequences in sections. These are named and defined. The section is concluded with a description of some general concepts.

Definitions 245–255 (Ancient Reef Morphology)

245. REEF (geological); REEF (stratigraphic): a general term for a (carbonate) body that is laterally restricted and locally formed. It has been developed with some topographic relief and occurs as local mounds that may be arranged in regional curvilinear trends (see Fig. 2–57).

NE rif
DE Riff
ES arrecife
FR récif
RE R

246. ORGANIC FRAMEWORK REEF (ECOLOGICAL REEF): build-ups formed in part by a wave-resistant framework constructed by organisms (see Fig. 2–58). This implies that such a "reef" exerts some degree of control over its surrounding environment. In geology, the terms have been used in the sense of

<table>
<tr><td colspan="2">Heckel (1974)
Compositional descriptive terminology</td><td colspan="3">Common usage followed in this text; descriptive with genetic implications</td><td>Dunham (1970)</td></tr>
<tr><td rowspan="4">Major mixed build–ups</td><td>Encrusted skeletal build–up</td><td>Organic framework reef</td><td rowspan="3">Bioherm</td><td rowspan="4">Carbonate build–up</td><td>Ecological reef</td></tr>
<tr><td>Loose skeletal build–up</td><td>Organic bank</td><td rowspan="3">Strati–graphic reef</td></tr>
<tr><td>Lime mud build–up</td><td>Lime mud accumulation</td></tr>
<tr><td>Sorted–abraded skeletal build–up</td><td>Sediment pile</td><td></td></tr>
</table>

Fig. 2–57. Terminology for carbonate build-ups as proposed by Wilson (1975), as opposed to Heckel (1974) and Dunham (1970) terms. Illustration after Wilson (1975), with permission of the author and Springer Verlag.

(1) organic framework communities and (2) organic build-ups. Consequently the terms, when used, should be modified.

Note: See Dunham (1970).

NE organisch rif
DE organische Riffe; organisches Riffgerüst
ES arrecife orgánico
FR récif organique
RE R

247. CARBONATE BUILD-UP: a body of locally formed, laterally restricted carbonate sediment which possesses topographic relief (see Fig. 2–59).

NE karbonaatopbouwsel
DE Karbonat-Hügelstruktur; Schlammhügel
ES construcción carbonatada
FR complexe carbonaté
RE P

248. CARBONATE MASS: a carbonate accumulation with only slight relief caused by facies changes from compactible argillaceous strata to noncompactible carbonates.

NE karbonaatmassa
DE Karbonatkörper
ES masa carbonatada
FR corps carbonaté
RE P

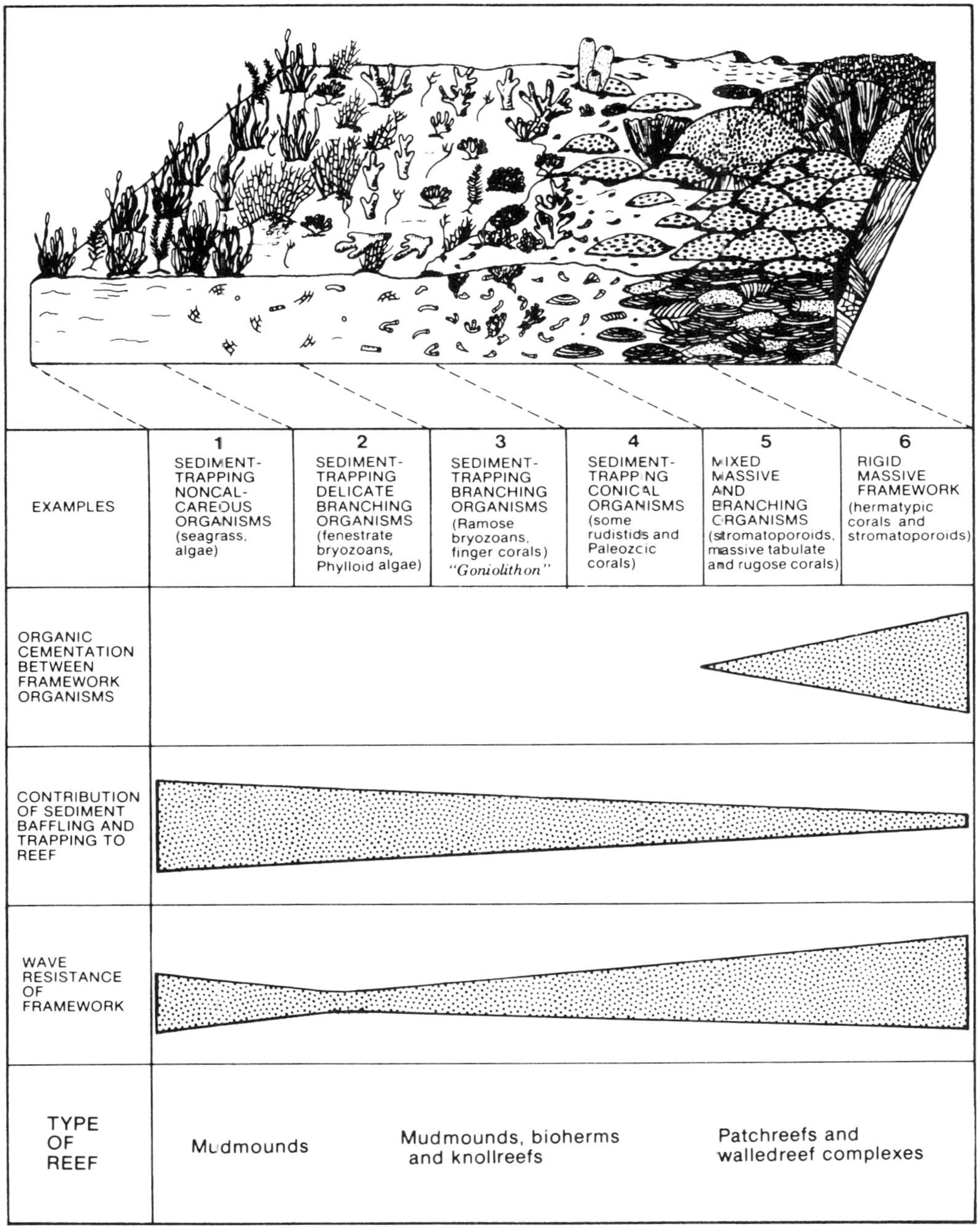

Fig. 2–58. Variations in reef framework and predominant resulting reef types. After Longman (1981). By permission of SEPM.

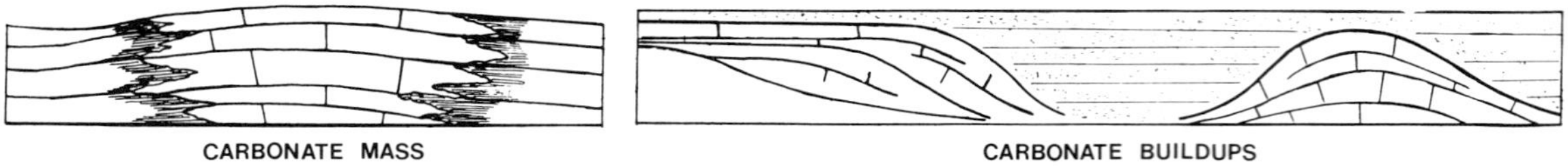

Fig. 2–59. Configurations of carbonate masses and carbonate build-ups. Illustration after Wilson (1975) by kind permission of the author and Springer Verlag.

249. MOUND: equidimensional or ellipsoidal build-up (see Fig. 2–60).

NE symmetrisch karbonaat opbouwsel
DE symmetrische Karbonat-Hügelstruktur
ES acumulacion symetrico carbonatado; "mound"
FR monticule
RE P

250. LIME MUD MOUNDS (LINEAR MUD ACCUMULATIONS): build-ups accumulated through both hydrodynamic processes and *in situ* organic production. Consequently a lime mud matrix dominates such constituents as organic boundstone and bioclastics (see Fig. 2–61).

NE modderige karbonaat opbouwsels
DE Kalkschlamm-Bioherm; stillwasser-Bioherm
ES "mounds" de fango carbonatado
FR monticule de boue
RE P

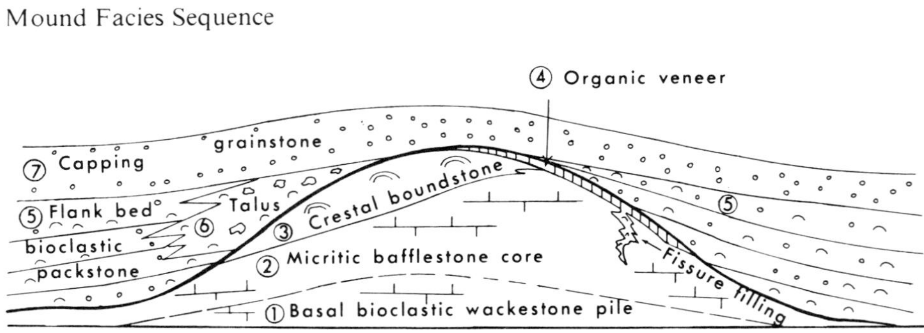

Fig. 2–60. Ideal carbonate mound with seven commonly developed facies. Variety in facies is the result of growth of the mound into the wave base. The facies are controlled by the rate of carbonate production, subsidence and hydrographic factors. Illustration by kind permission of Wilson (1975) and Springer Verlag.

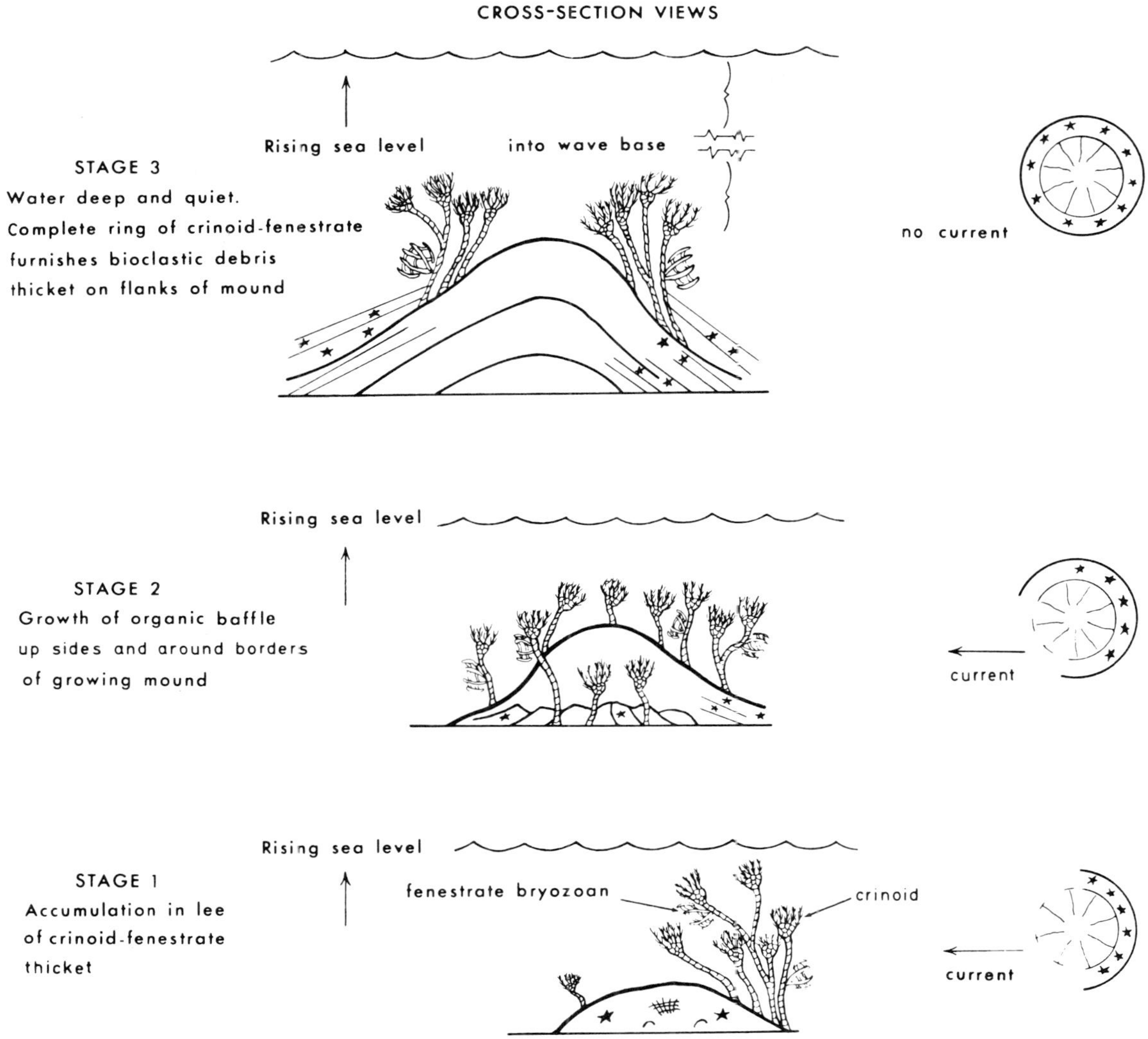

Fig. 2–61. Hypotheses of development of equidimensional Waulsortian mounds through progressive colonization of current-induced pile of fine sediment in lee of crinoid–fenestrate bryozoan thicket. Mud accumulates in centre of thicket, growing on all sides of mound. Flanking beds develop coarse debris derived from other edges of thickets where finest lime mud is winnowed. Illustration by kind permission of Wilson (1975) and Springer Verlag.

251. BIOHERM: reeflike, moundlike, lenslike or otherwise circumscribed structure of strictly organic origin, embedded in rocks of a different lithology. Wilson (1975) suggested that it refers to a build-up whose internal composition shows it to be derived largely from *in situ* production of organisms or as a framework of encrusting growth as opposed to mainly mechanical (hydrodynamical) piling.

NE bioherm
DE Bioherm
ES bioherma; biohermo
FR bioherme
RE R

252. (REEF) KNOLL (KNOLL REEF): a bioherm of organic frame-building organisms, represented by a small, isolated, prominent, rounded hill up to 100 m high, consisting of resistant reef material; also used for Recent reefs, which are then isolated, more or less circular, in shelf-marginal or basinal settings.

NE rifknobbel; knollenrif
DE Knollenriff
ES pináculo arrecifal
FR intumescence récifale
RE R

253. BIOSTROME: a bedded or layered structure, platelike and not moundlike, built mainly by sedentary organisms and composed largely or exclusively of the same.

NE biostroom
DE Biostrom
ES biostroma
FR biostrome
RE Q

254. BIOLITHITE: a limestone constructed by organisms that grew and remained in place. It is characterised by a rigid framework of carbonate material that binds allochems and bioclasts (e.g., reef core). The major organism should be specified when the term is used (e.g., coral biolithite).

NE organogeen, biogeen gesteente
DE Biolithit; autochthones Riffgestein
ES biolitito
FR biolithite
RE RLS

255. BIOLITH: biogenic rock, organic rock; deposits due directly to the physiological activities of organisms.

NE biogeen, organogeen gesteente; bioliet
DE Biolith
ES biolito
FR biolite; biolithe
RE RLS

Definitions 256–262 (Sedimentary Mechanisms, Concepts and Sequences)

256. CARBONATE SEDIMENTARY MECHANISM: the accumulation of carbonate masses as reflected in their facies distribution due to a response to the dynamic equilibrium carbonate production or destruction and sea level changes or bottom movements (see Figs. 2–62 and 2–63).

NE karbonaat sedimentair mechanisme
DE Karbonatsedimentationsmechanismus
ES mecanismo sedimentario carbonatado
FR mécanisme sédimentaire carbonaté

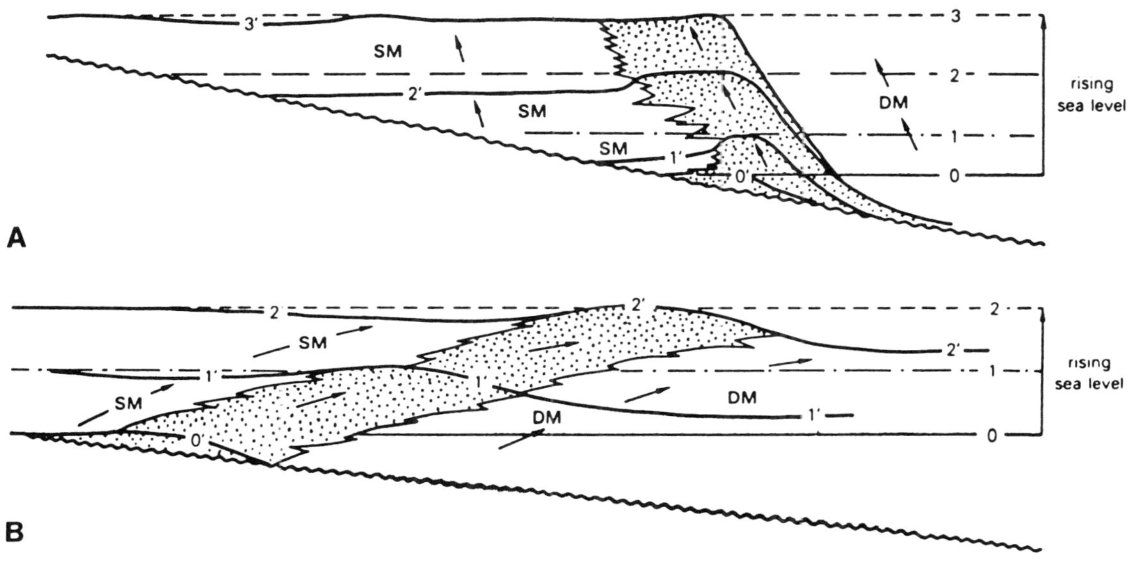

Fig. 2–62. Schematic representation of different sedimentation models on smooth foundations resulting from balance between carbonate production and subsidence. A, Stationary model (foundation attached to landmass); B, progradation model (foundation attached to landmass). Key 2′, sediment–water interface at sea level 2; →, accretion direction; SM, shallow marine carbonaceous mud; DM, deeper marine carbonaceous mud; stippled areas, oölitic and skeletal sands, reefs.

Carbonate production in balance with depth variations		Carbonate production not in balance with depth variation	
	CP = DV	CP < DV	CP > DV
Depth decreasing*	Erosion	Erosion	Carbonate mass builds out
Depth constant*	Non–deposition	Erosion/ non–deposition	Carbonate mass builds out
Depth increasing*	Carbonate mass builds up	Carbonate mass builds in or is drowned (goes beyond zone of carbonate production)	Carbonate mass builds out

Fig. 2–63. Sedimentation patterns with a constant carbonate production (CP) in an environment with various depths (DV). The asterisks indicate depth fluctuations resulting from sea level fluctuations or sea bottom movements.

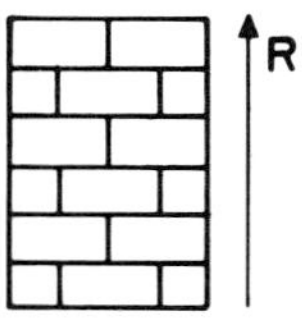

257. REGRESSIVE "REEF" (SEQUENCE): a sequence of nearshore reefs or bioherms superimposed on basinal deposits and covered by lagoonal–supratidal sediments deposited behind younger nearshore reefs during the rising of a landmass or the lowering of the sea level, and developed more or less parallel to the shore (see Fig. 2–64).

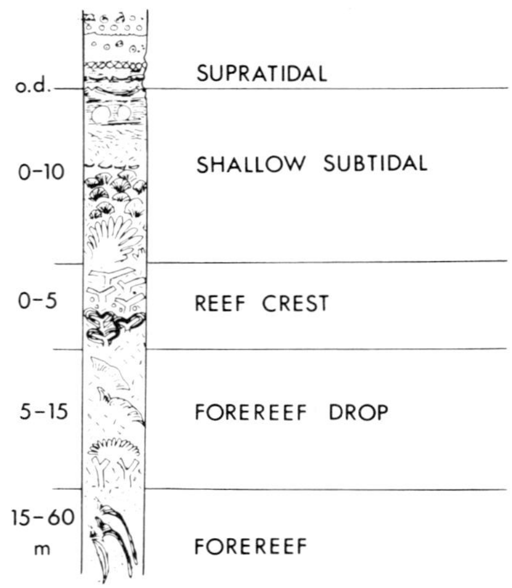

Fig. 2–64. Schematic section through a regressive cycle in a Recent–Pleistocene coral–algal reef. Illustration by permission of Dr. J.A.F.B. Hubbard and Nottingham University Press.

NE regressief rif
DE regressives Riff
ES arrecife regresivo
FR récif de regression

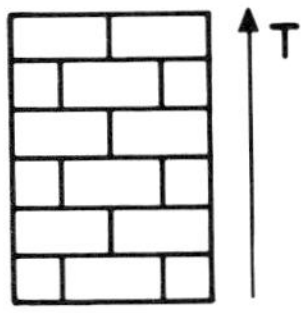

258. TRANSGRESSIVE "REEF" (SEQUENCE): a sequence of nearshore reefs or bioherms superimposed on backreef deposits of older reefs during sinking of a landmass or rising of the sea level, and developed more or less parallel to the shore (see Fig. 2–65).

NE transgressief rif
DE transgressives Riff
ES arrecife transgresivo
FR récif de transgression

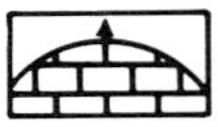

259. BUILD-UP CARBONATE BODY: carbonate body (mass) formed as a response to positive sea level changes or subsidence of the sea floor; the carbonate production should keep pace with the movements (see Fig. 2–66).

NE opgebouwd karbonaat lichaam
DE Hügelstruktur-Karbonatkörper; Hügelstrukturierter Karbonatkörper, Karbonatplattform
ES cuerpo de construcción carbonatada
FR corps carbonaté construit
RE P

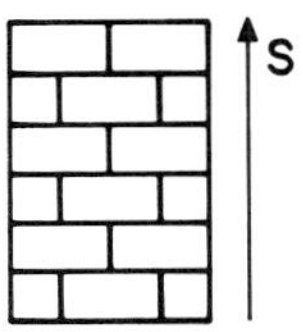

260. SHOALING-UPWARDS CYCLE: a sedimentation mechanism, being the result of carbonates accumulating faster than positive sea level movements or negative sea bottom movements and resulting in a succession of lithofacies that, in an upward direction, reflect increasingly shallower environmental conditions (see Fig. 2–67).

NE ondieper wordende opeenvolging
DE Sedimentation in zunehmend flacheren Bereichen
ES ciclo de disminución de profundidad hacia arriba
FR cycle de diminution de profondeur

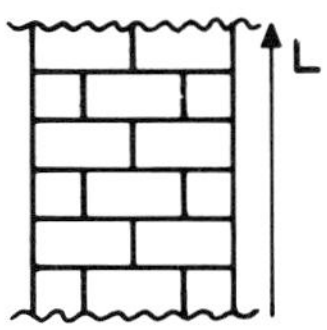

261. LOFER (LOFERITE) CYCLE: a carbonate sequence characterised by a succession in an upward direction of a basal, argillaceous or conglomeratic member representing reworked weathering material, an intertidal unit with algal mats, fenestral fabrics and abundant dissication features and a shallow marine limestone with shallow marine organisms. Finally, a drop in sea level produces another disconformity and a phase of intense weathering diagenesis (see Fig. 2–68).

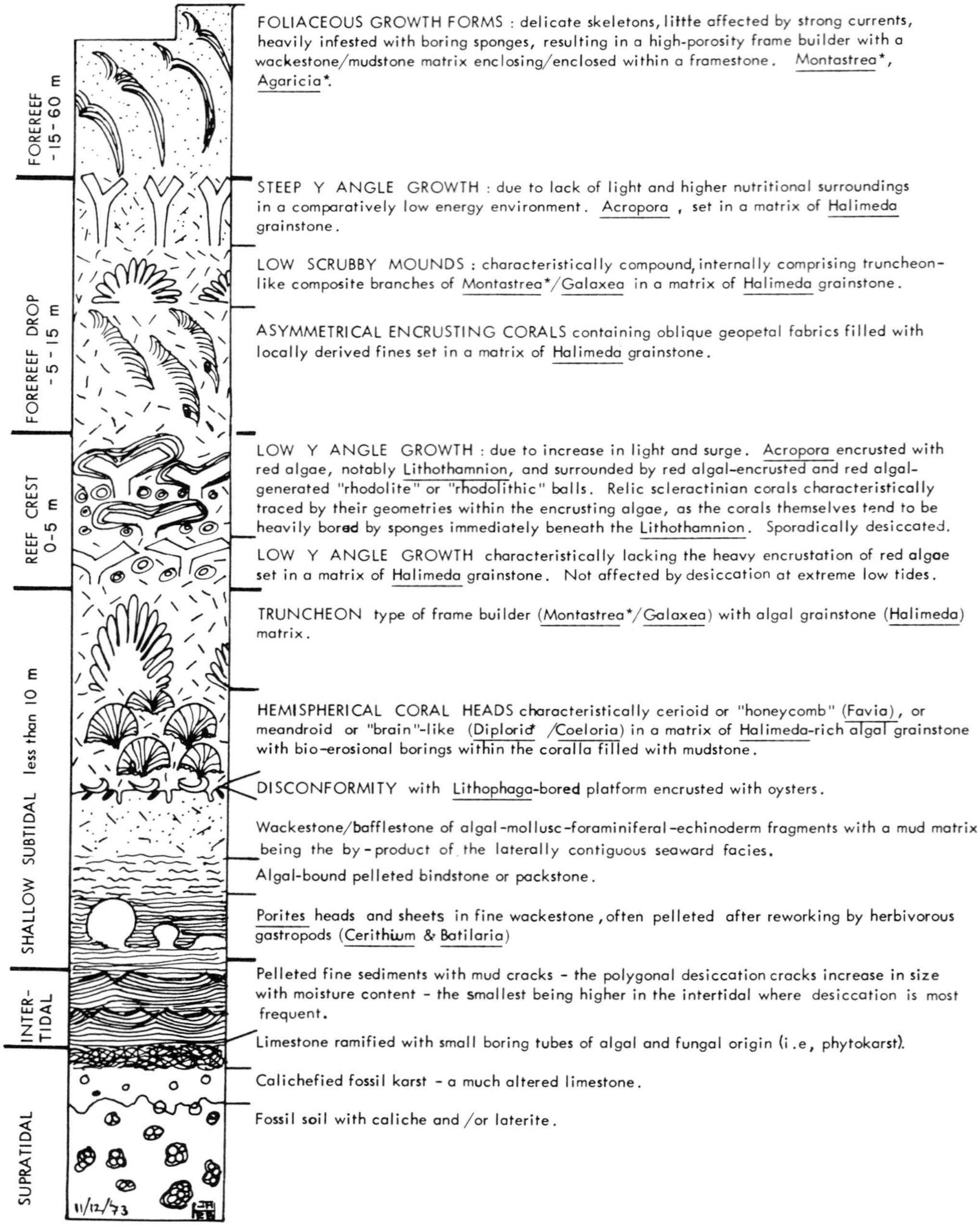

Fig. 2–65. Schematic section through a transgressive cycle in a Recent–Pleistocene coral–algal reef. Asterisks denote coral morphological counterparts from the Atlantic. Illustration by permission of Dr. J.A.E.B. Hubbard and Elsevier Publishing Company.

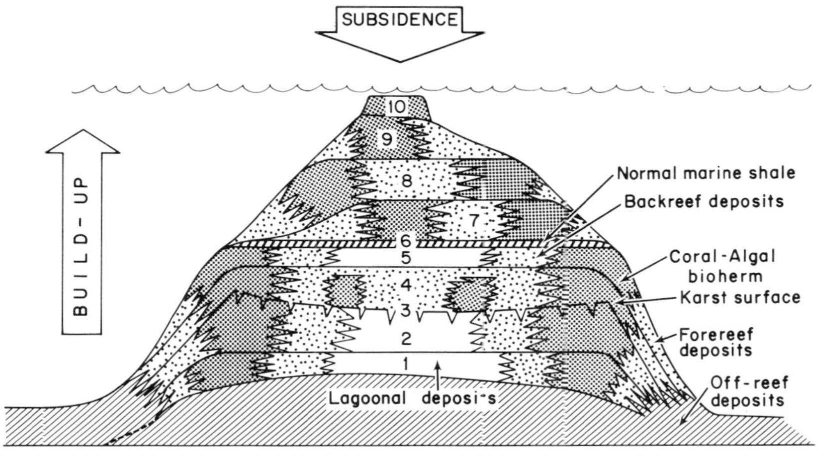

Fig. 2–66. Growth of a coral–algal reef complex (a build-up) in response to variations in sea level. Key: 1–10, successively younger events; 1, 2, 5, large lagoons without patch reefs; 4, large lagoon with patch reef; 7, 8, small lagoons filled with reef debris; 9, open-ring reef; 10, coral knoll; 3, erosion phase; 6, drowning phase.

NE loferopeenvolging
DE Loferitzyklen
ES loferita; ciclo loferítico
FR cycle de type lofërite

262. INTERNAL ARCHITECTURE: the internal variation in distribution of characteristic facies as a result of sea level changes during the formation of an actively growing carbonate mass (see Figs. 2–69 to 2–74).

NE interne opbouw
DE Internstruktur
ES distribución interna de facies
FR architecture interne

Definitions 263–268 (Miscellaneous Reef Concepts)

263. CONCEPTUAL DEPOSITIONAL MODEL: a reflection of depositional environments prevailing over a sufficient time span in a certain area to produce a recognisable imprint of the environmental conditions on the sediments. It is thus a model of depositional facies patterns and aids in predicting lateral facies changes (see Figs. 2–73 and 2–74).

NE afzettingsmodel
DE Konzeptuelles Ablagerungsmodell

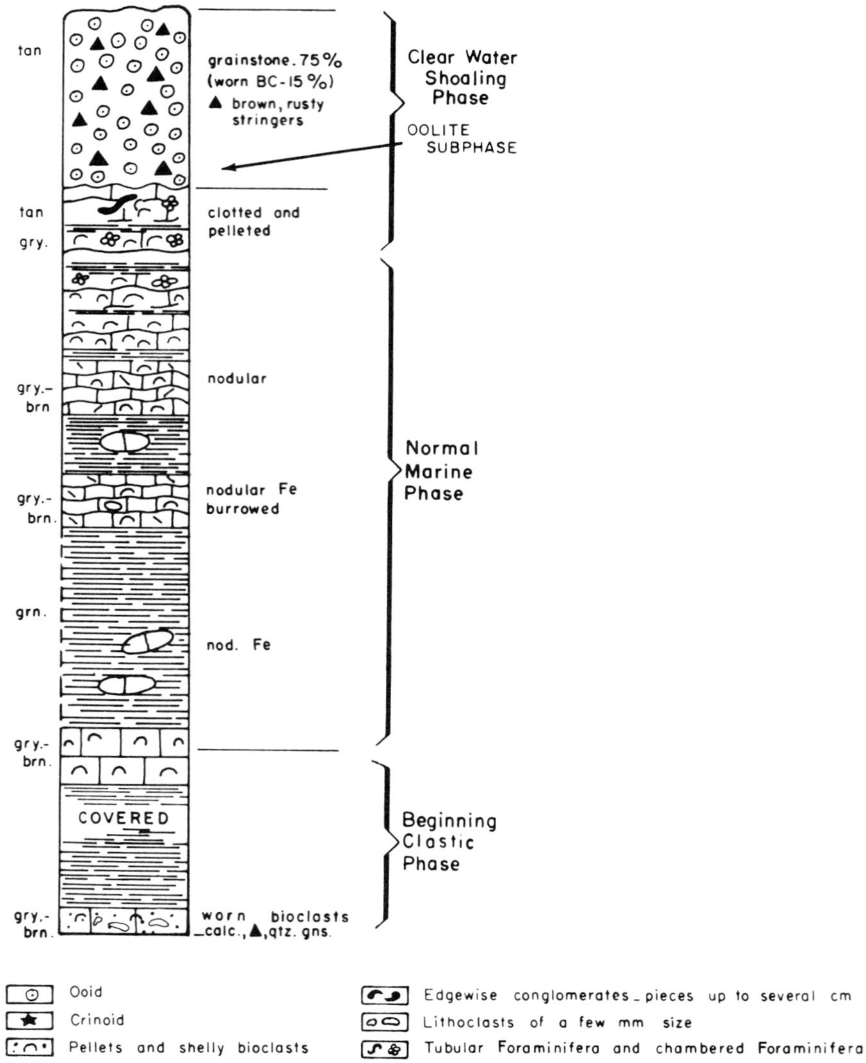

Fig. 2–67. (Left) Upward shoaling cycle; approximately 13-m measured section. Thick lines are contacts between cycles. They are probably marine hardgrounds because in the oolite there is no evidence of subaerial cementation. Timber Creek, Big Snowy Mountains, Montana, Mississippian Lodgepole Formation. Illustration by kind permission of Dr. J.L. Wilson, 1975, and by Springer Verlag. (Right) Ten-meter-thick cycle with algal plate mounds developed in clotted–pelleted foraminiferal strata just beneath an oolitic interval that reflects a shoaling sequence. Beeman Canyon, Sacramento Mountains, New Mexico. Illustration by kind permission of Wilson (1975), and by Springer Verlag.

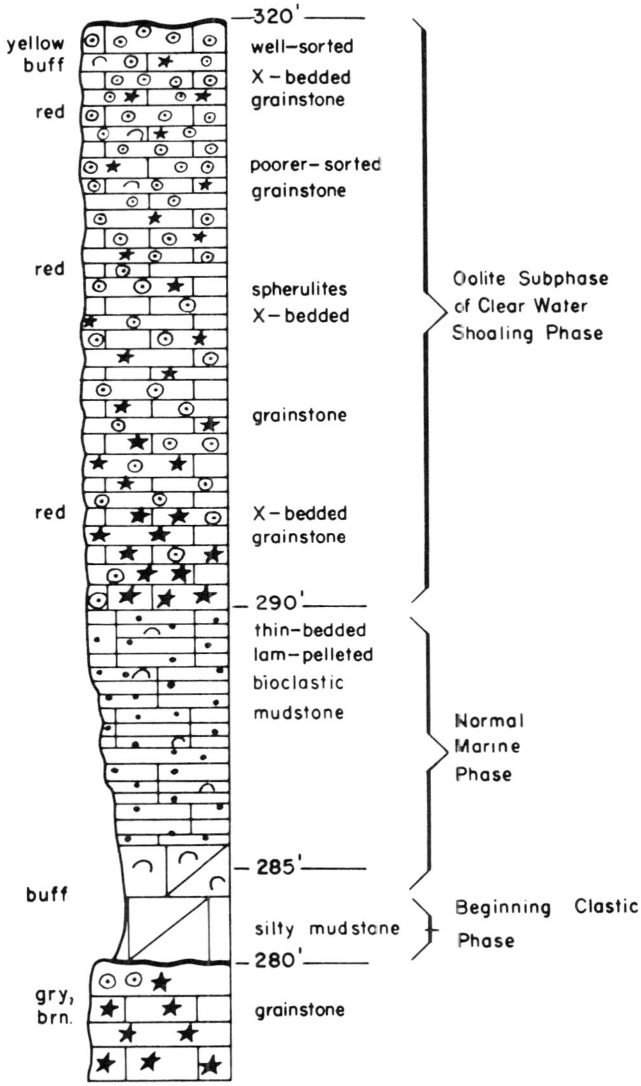

Fig. 2–67. *(Continued)*

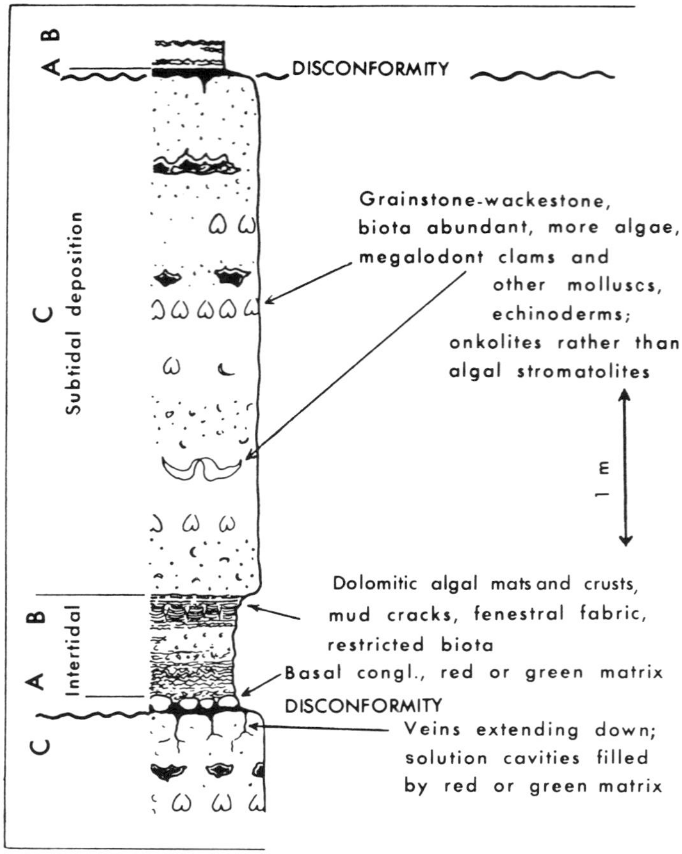

Fig. 2–68. Diagrammatic representation of a Lofer cyclothem. A, Basal, argillaceous member, commonly reworked residue. B, Intertidal member with algal mats and dessiccation features. C, Shallow marine "megalodont limestone" with dessiccation cavities caused by solution of subsequent sea level drop. Illustration by permission of Springer-Verlag.

TYPE OF REEF COMPLEX	AREA (km^2)	% OF BIOHERMAL LIMESTONE	% OF REEF DEBRIS DEFOSITS	% OF LAGOONAL DEPOSITS
REEF KNOLLS* (e.g.,Yucatan shelf)	< 1	90	5 – 10	NONE
REEF WALLS & SMALL OPEN-RING REEFS* (e.g,Arenase, Arcas reefs ,Yucatan)	2–5	60–80	20–40	NONE
CLOSED-RING REEFS* (e.g., Alacran reef , Yucatan)	250	5–10	5–10	80–90
LARGE CLOSED-RING REEFS (e.g,Eniwetok Atoll, Midway Islands)	750	<(?)5	<(?)5	>(?)90

A

Fig. 2–69. A, Average composition of Recent coral–algal build-ups in terms of biohermal limestone, reef debris deposits and lagoonal deposits. Data for complexes marked with an asterisk are taken mainly from Logan *et al.* (1969). B, Summary of the main features of Recent coral–algal reef complexes.

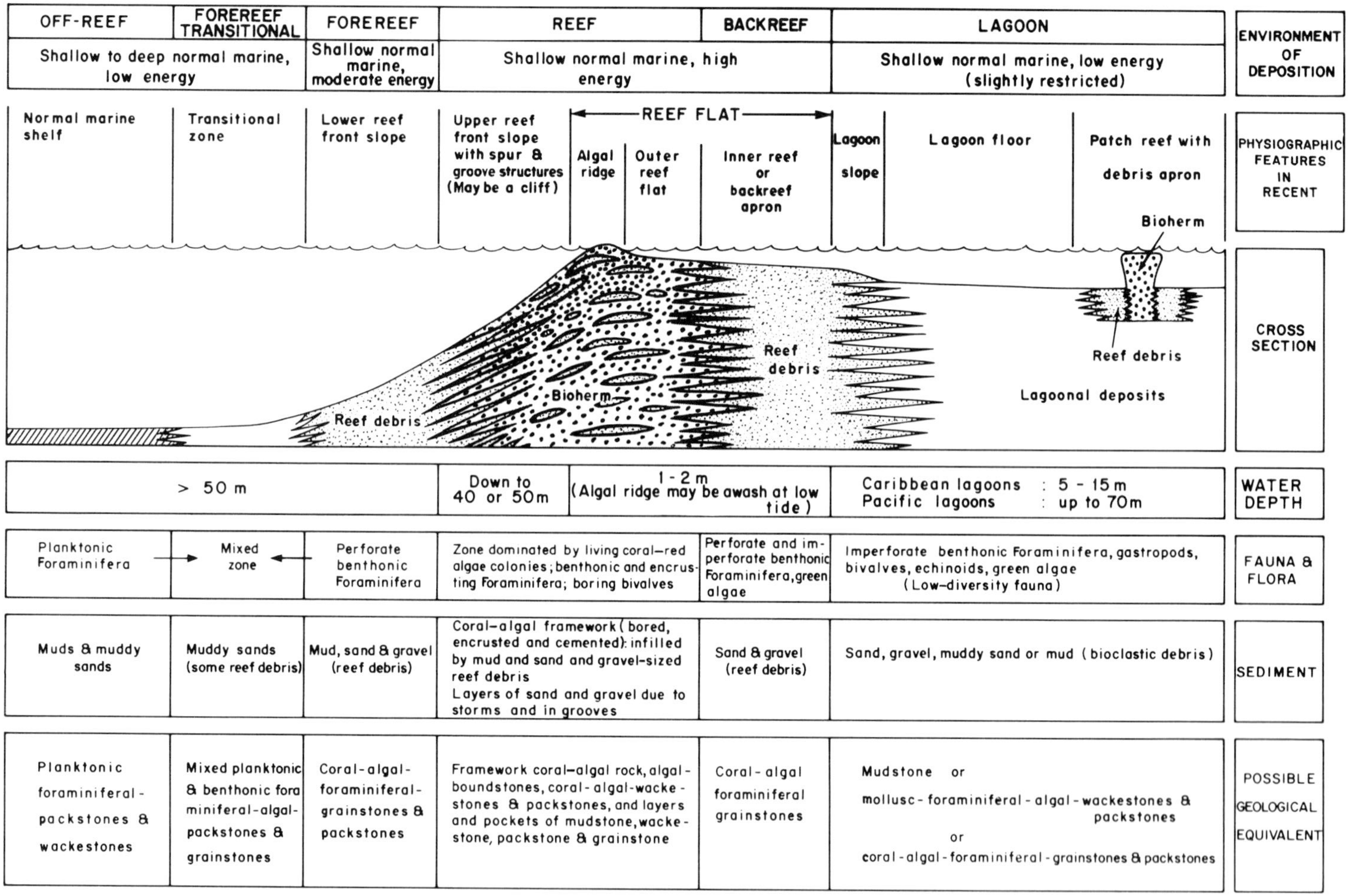

B

Fig. 2–69. *(Continued)*

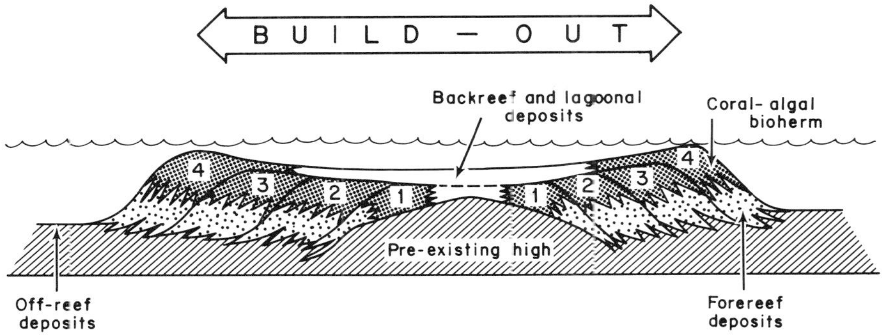

Fig. 2–70. Growth of a coral–algal reef complex in a lateral direction (build-out) at more or less constant sea level. The numbers 1 to 4 indicate successively younger growth phases.

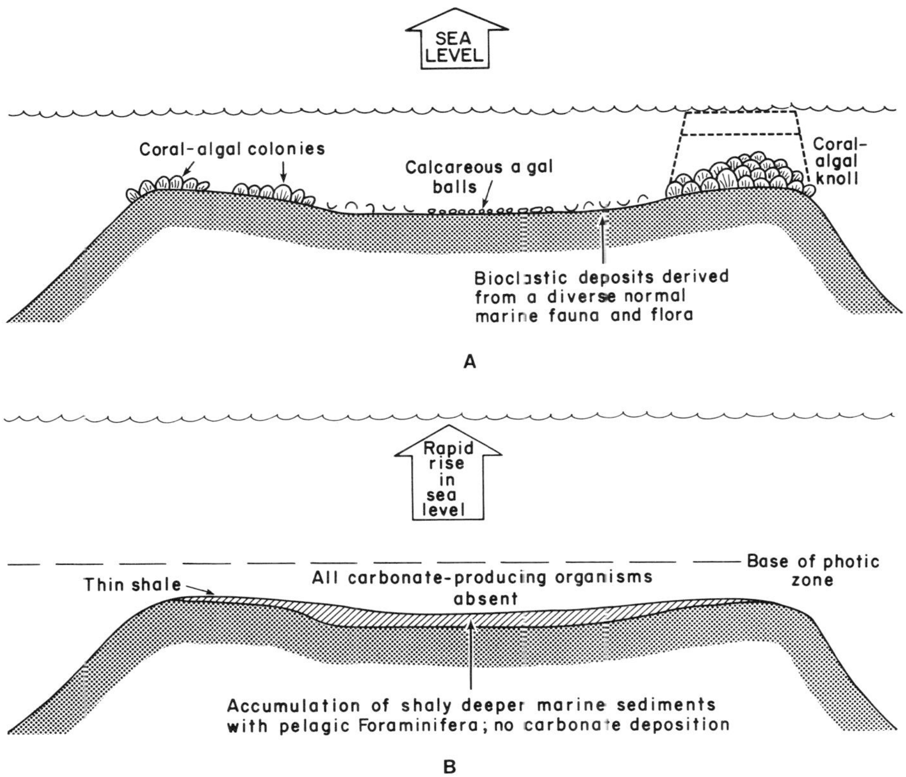

Fig. 2–71. A, Submerged carbonate banks are the result of sea level rising faster than the rate of sediment accumulation or carbonate production. Coral–algal reef complex becomes submerged, and a shallow marine bank environment is established; carbonate production continues. B, Drowned carbonate banks or reef complexes are the result of a rapid rise in sea level surpassing the rate of carbonate production. This carbonate production virtually stops in hermatypic coral colonies below the photic zone.

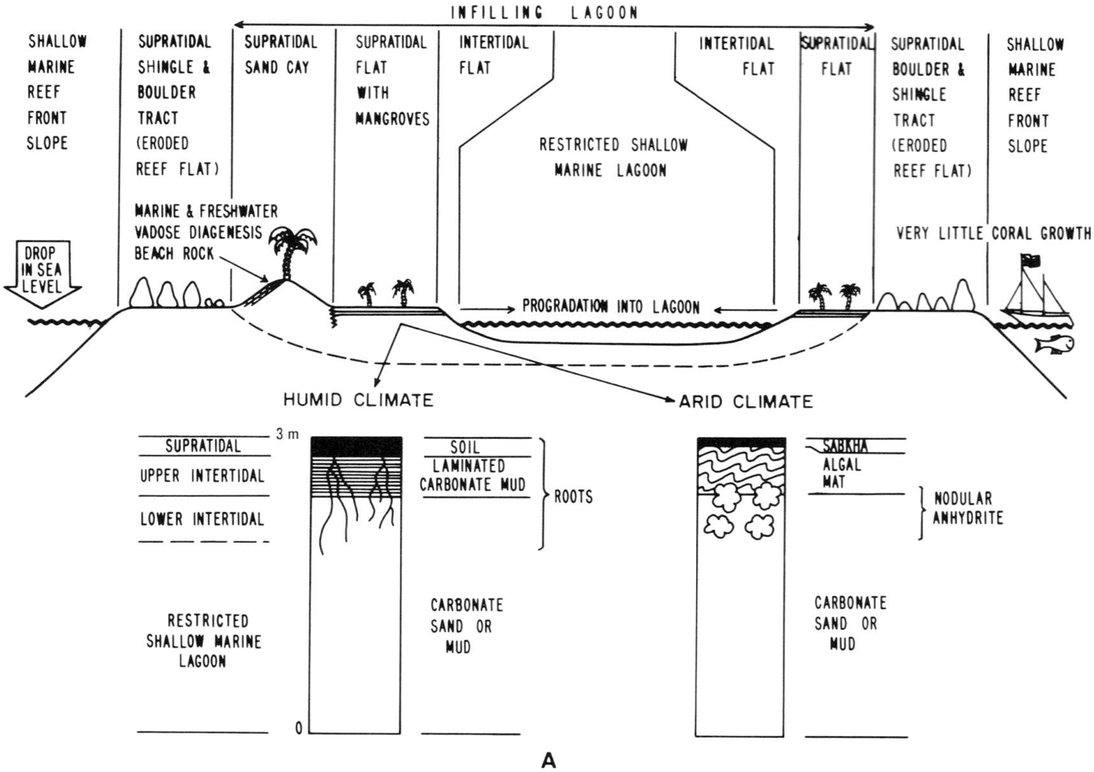

Fig. 2–72. A, A slight drop in sea level could result in an inactive reef complex in which in the supratidal flat characteristic sedimentary sequences develop in humid and in arid climates. B, A large drop in sea level exposes the entire reef complex. It is subject to karstification. Solution weathering by fresh water results in a solution rim above which later (after rise in sea level) new reefs could preferentially form.

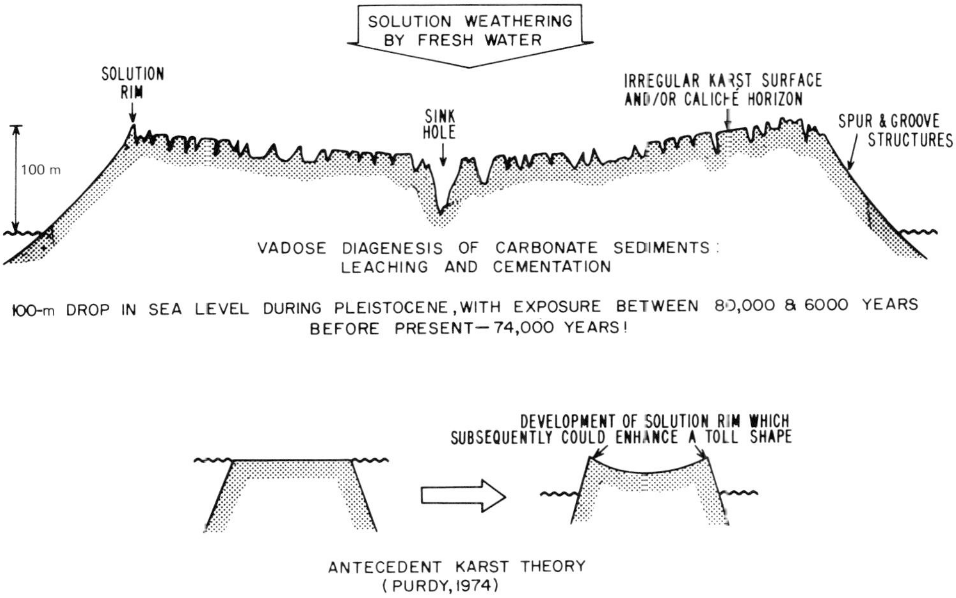

B

Fig. 2–72. (*Continued*)

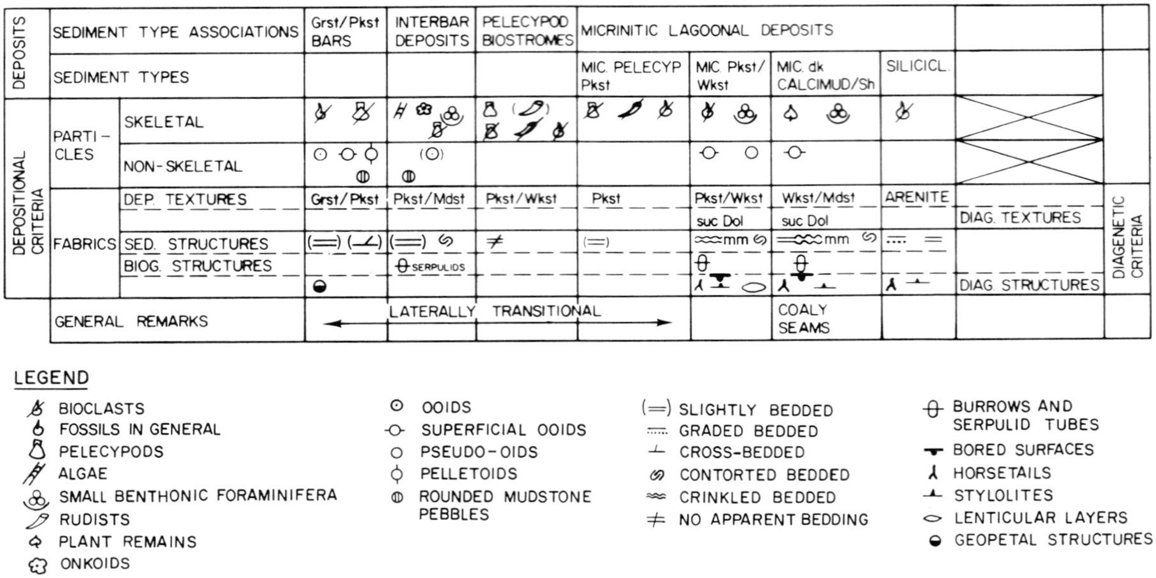

Fig. 2–73. Example of main depositional and diagenetic criteria that enable one to formulate a meaningful depositional model (Cretaceous Cogollo Group, Maracaibo Basin, Venezuela). Grst, grainstone; Pkst, packstone; Wkst, wackestone; Mdst, mudstone; suc. Dol, sucrosic Dolomite; Mic, micrimite; Sh, shale; dk, dark.

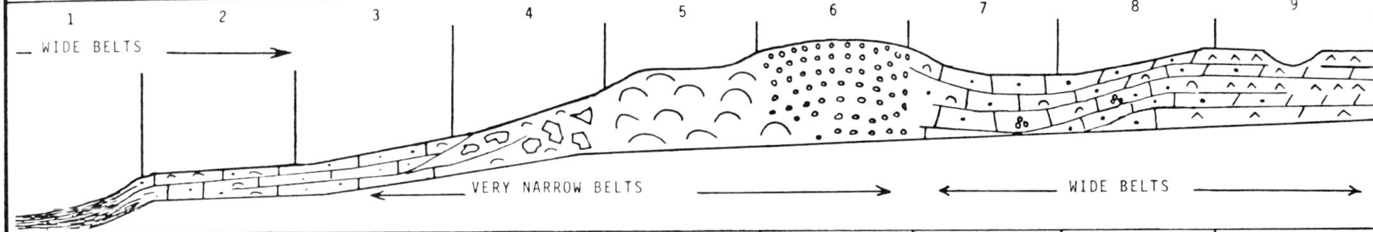

	BASIN	OPEN SEA SHELF	DEEP SHELF MARGIN	FORESLOPE	ORGANIC BUILD-UP	WINNOWED EDGE SANDS	SHELF LAGOON OPEN CIRCULATION	RESTRICTED CIRCULATION SHELF AND TIDAL FLATS	EVAPORITES ON SABKHAS-SALINAS
FACIES PROFILE	1	2	3	4	5	6	7	8	9
2ND-ORDER SEDIMENTARY BODIES			DEBRIS FLOWS AND TURBIDITES IN FINE LAMINATE STRATA, MOUNDS ON TOE OF SLOPE	GIANT TALUS BLOCKS, INFILLED LARGE CAVITIES, DOWNSLOPE MOUNDS	DOWNSLOPE MOUNDS, REEF KNOLLS, BOUNDSTONE PATCHES, FRINGING AND BARRIER FRAME-WORK REEF, SPUR AND GROOVE	ISLANDS, DUNES, BARRIER BARS, PASSES AND CHANNELS	TIDAL DELTAS, LAGOONAL PONDS, TYPICAL SHELF MOUNDS, COLUMNAR ALGAL MATS, CHANNELS AND TIDAL BARS OF LIME SAND	TIDAL FLATS, CHANNELS, NATURAL LEVEES, PONDS, ALGAL MAT BELTS	ANHYDRITE DOMES, TEPEE STRUCTURES, LAMINATED CRUSTS OF GYPSUM, SALINAS (EVAPORATIVE PONDS), SABKHAS (EVAPORATIVE FLATS)
STANDARD MICROFACIES	1 SPICULITE 2 MICROBIO-CLASTIC CALCISILT 3 PELAGIC MICRITE RADIOLARITE SHALE	2 MICROBIOCLASTIC CALCISILT 8 WHOLE SHELLS IN MICRITE 9 BIOCLASTIC WACKESTONE 10 COATED GRAINS IN MICRITE	2 MICROBIOCLASTIC CALCISILT 3 PELAGIC MICRITE 4 BIOCLASTIC-LITHOCLASTIC MICROBRECCIA	4 BIOCLASTIC LITHOCLASTIC MICROBRECCIA, LITHOCLASTIC CONGLOMERATE 5 BIOCLASTIC GRAINSTONE-PACKSTONE, FLOATSTONE 6 REEF RUDSTONE	7 BOUNDSTONE 11 COATED, WORN, BIOCLASTIC GRAINSTONE 12 COQUINA (SHELL HASH)	11 COATED, WORN BIOCLASTIC GRAINSTONE 12 COQUINA (SHELL HASH) 13 ONKOIDAL BIOCLASTIC GRAINSTONE 14 LAG BRECCIA 15 OOLITE	8 WHOLE SHELLS IN MICRITE 9 BIOCLASTIC WACKESTONE 10 COATED GRAINS IN MICRITE 16 PELSPARITE 17 GRAPESTONE ONKOIDS IN MICRITE 18 FORAM, DASYCLADACEAN GRAINSTONE	16, 17, 18 19 FENESTRAL, PELOIDAL, LAMINATE, MICRITE 24 RUDSTONE IN CHANNELS 21 SPONGIOSTROME, MICRITE 23 NON LAMINATE, PURE MICRITE 22 ONKOIDAL MICRITE	20 STROMATOLITIC MICRITE 23 NON-LAMINATE, PURE MICRITE, NODULAR-PEARL ENTEROLITHIC ANHYDRITE, SELENITE BLADES IN MICRITE

Fig. 2–74. Synopsis of standard facies belts (Wilson, 1975) reviewing second-order sediment bodies and standard microfacies associated with each level. The numbers 1–23 are Wilson's standard microfacies. Illustration by kind permission of Dr. J.L. Wilson and Springer-Verlag.

ES modelo deposicional
FR modèle conceptuel de dépôt

264. DYNAMIC REEF CONCEPT: reefs understood as the site and the product of a dynamic interplay of construction and destruction by organisms, sedimentation, cementation and mechanical breakdown, commonly resulting in packstones and wackestones with a low porosity.

NE dynamisch rif concept
DE Dynamisches Riffkonzept
ES concepto dinámico de arrecife
FR concept de dynamique récifale

265. STATIC REEF CONCEPT: reefs understood as the site and the product of a static but commonly rapid growth of framework builders, essentially unaffected by destructive modifiers and resulting in easily recognisable biohermal limestone with good porosity.

NE statisch rif concept
DE Statisches Riffkonzept
ES concepto estático de arrecife
FR concept de statique récifale

266. REEF CALCIUM CARBONATE BUDGET: the path of calcium carbonate through the reef ecosystem expressed in net loss and net gain in volumetric terms over a fixed period of time (see Fig. 2–75).

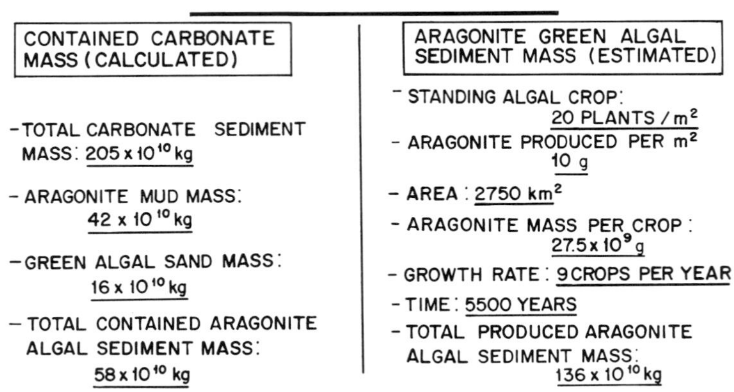

Fig. 2–75. Green algal carbonate budget bight of Abaco. After Neumann and Land (1975).

Note: See Stearn *et al.* (1977).

NE $CaCO_3$ budget van een rif
DE $CaCO_3$ Haushalt im Riff
ES cómputo del $CaCO_3$ de un ecosistema arrecifal
FR bilan du carbonate de calcium récifal

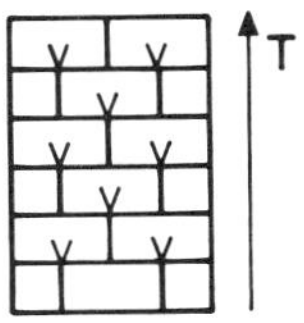

267. TRANSGRESSIVE TIDAL FLAT: a tidal flat that is characterised by stages in its development demonstrating active transgression of branching tidal channels dissecting the tidal flats.

NE transgressieve getijdenvlakte
DE Transgressives Watt
ES llanura mareal de tipo transgressivo
FR slikke sableuse de transgression
RE CT

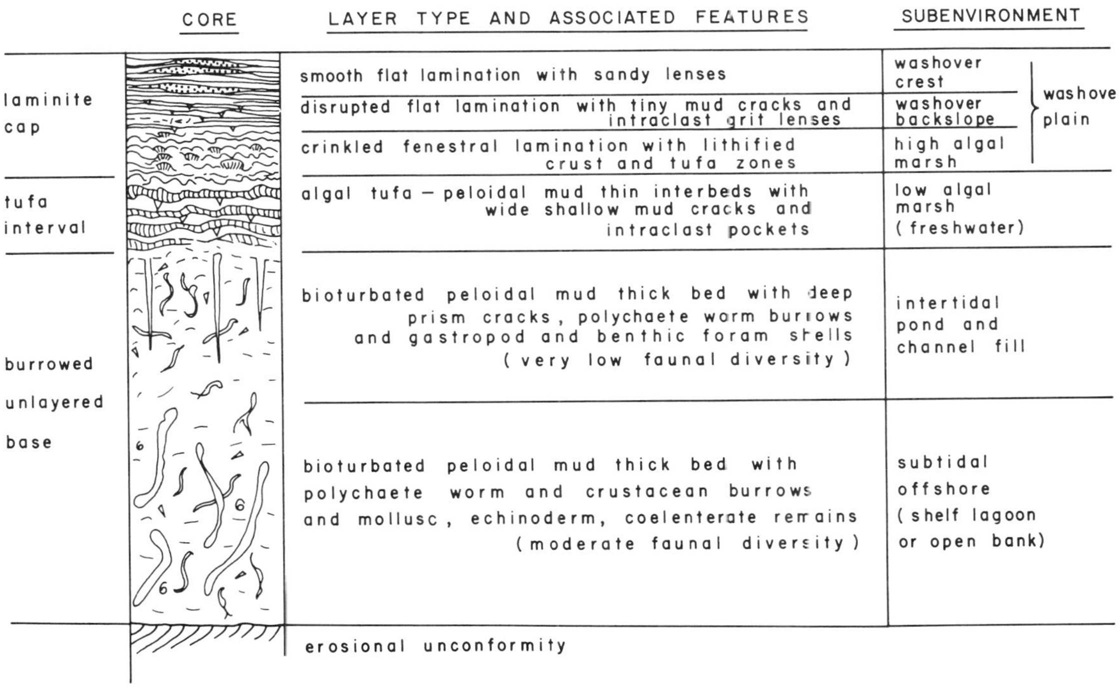

Fig. 2–76. Schematic drawing of a vertical section through Andros tidal flats (Bahamas) at the completion of a regressive cycle. Illustration after Hardie (1977). By permission of the John Hopkins University Press.

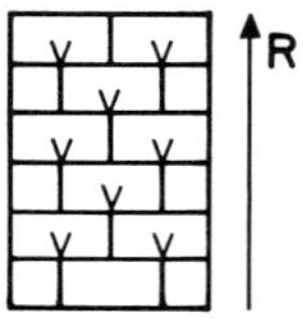

268. REGRESSIVE TIDAL FLAT: a tidal flat that is characterised by stages in its development demonstrating active regression. Features to demonstrate this are the development of broad exposed coastal plains and the presence of tidal ponds or lakes in such plains. Moreover, scars of abandoned tidal channels might be present (see Fig. 2–76).

NE regressieve getijdenvlakte
DE regressives Watt
ES llanura mareal de tipo regressivo
FR slikke salsleuse de régression
RE CT

3

Carbonate Diagenesis

3–1 INTRODUCTION

Diagenesis is continuous. In rocks, it leaves series of marks. If broken down in steps, diagenesis is understood to work repeatedly as a process that affects a carbonate component or a carbonate mass, resulting in one or more of the following features: cements, diagenetic textures or structures, diagenetically formed minerals or a secondary pore. Diagenetic processes act in various depth or time settings, as a result of which identical or different products could form that modify the rock. Thus associations of diagenetic features can characterise time–depth settings. With this in mind the following overall subdivision has been made and will serve as a guide for further discussion:

1. Aspects of diagenesis
2. Diagenetic processes
3. Diagenetic products

In the first part a "frame of thought" is presented and general terms are defined. In the second and third parts specific processes and products in diagenesis are discussed. Although it could be argued that a process and its ensuing product should be defined together, the more fundamental and frequently overlooked distinction between process and product has been chosen here as the basis for a first-order subdivision. Cross references refer processes to products and vice versa.

3–2 ASPECTS OF DIAGENESIS

Introduction

The word *diagenesis* was applied for the first time by von Gümbel (1888), and the concept has since been used and misused by scores of geologists and petrologists. Most present-day workers have basically maintained the definition formulated by Walther (1894): "By diagenesis we mean all the physical and chemical changes that a rock undergoes from the time of its deposition, other than changes due to the intervention of tectonic compression and volcanic heat''.

Dunoyer de Segonzac (1968) pointed out that a lack of interchange of ideas developed and published in various languages has led to peculiar differences of interpretation of Walther's definition. These differences tend to concentrate on the boundaries of the field of diagenesis, the time interval encompassed by diagenetic processes, the zones in which particular diagenetic processes act and the nature of the diagenetic processes. These factors will be successively discussed.

Boundaries of the Field of Diagenesis

In nature, the boundaries of any field in the lithogenetic cycle are vague. In defining the field of diagenesis, one must ask the following questions:

1. When does diagenesis start?
2. At what degree of transformation of the original rock (and how measured) should the term *diagenesis* be replaced by the term *metamorphism?*
3. Are surface-weathering phenomena diagenetic?

The answers to these questions (see, e.g., Dunoyer de Segonzac, 1968) are not only academic but also interesting from an oil geological point of view. In the area of oil geology, approaches are followed that lead to delineating host or reservoir rock trends that can be extrapolated into as yet uninvestigated areas, and reservoir quality is considerably controlled by diagenesis. Therefore, the three questions formulated above merit further discussion.

The moment in which diagenetic processes start appears to be well described in Walther's definition: diagenesis starts ''from the time of [a sediments's] deposition''. But this still leaves some question as to the status of such processes as ooid–pisoid coating of grains, algal activity (micrite-rim formation) and lithoclast formation. Walther's definition of diagenesis excludes such processes, but it is thought that they are not exclusively sedimentary. This could be resolved by the inclusion of such processes in a group of processes that occupies a position intermediate between sedimentation and diagenesis. Soil forming also takes place within this group of processes.

The question as to when metamorphic processes take over diagenetic ones has been answered adequately in Winkler's (1967) paper on this matter. His principal criterion for the diagenetic field is the retention of sedimentary structures and the presence of mineral associations that characterise depositional pressure–temperature conditions.

Time Interval During Which Diagenetic Processes Act

Diagenesis starts from the moment of final deposition of a sediment. Usually it is difficult to indicate over which time interval a diagenetic process acts. Purdy (1968) stated that the degree of diagenetic alteration is related to the duration of exposure to a post-depositional environment. In individual studies and after calibration, Purdy's concept might give a relative scale of comparison. This is illustrated in Fig. 3–1 for a hypothetical case. In many diagenetic studies, however, it is impossible to pinpoint exactly when a diagenetic change takes place. However, pinpointing this moment can be important, particularly when the change modifies a reservoir. It aids in evaluating the chance that reservoir rocks are charged with hydrocarbons if the maturation of source rock and migration of hydrocarbons can also be pinpointed in time.

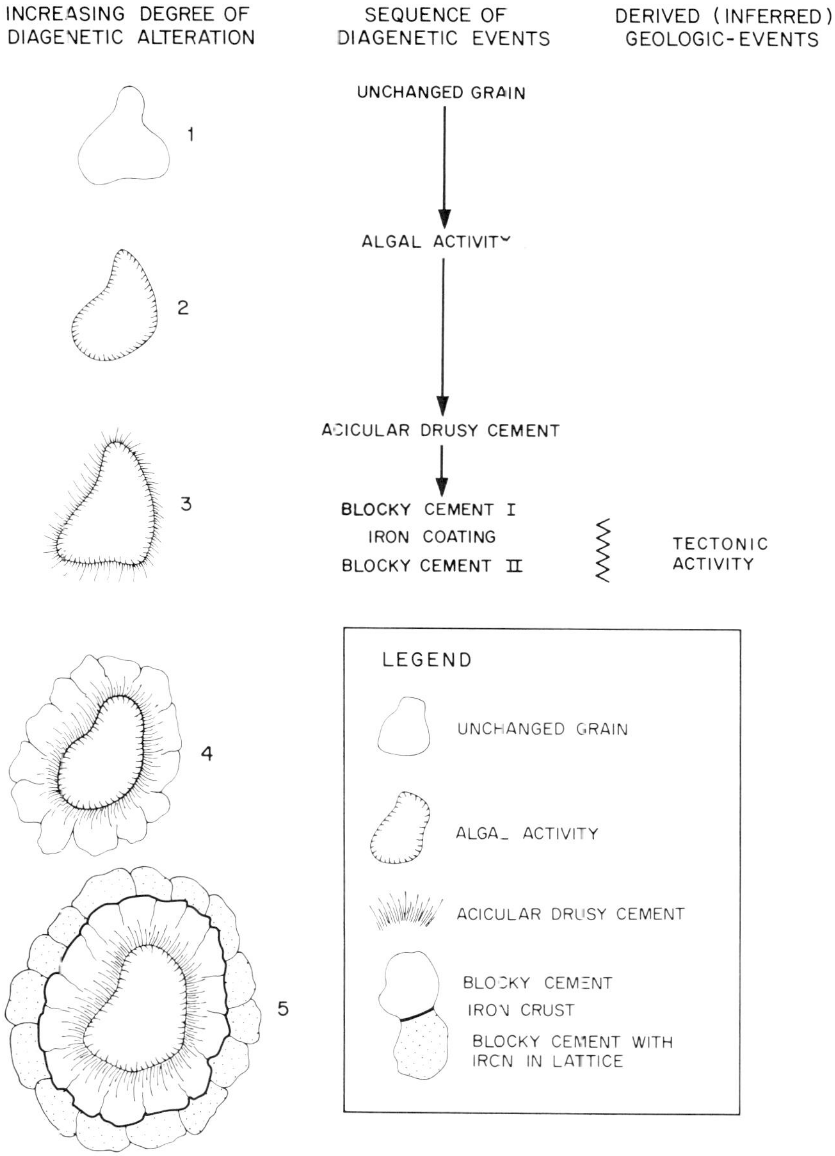

Fig. 3–1. Diagenetic evolution of a particle and derived (inferred) sequence of diagenetic events. Carbonate grains may have a fine drusy acicular rim of cement crystals, followed by a coarser blocky cement that fills part of the remaining pore space. In such a case three time intervals are recognisable: the time before cement started to precipitate, the period in which drusy cement crystals rimmed the grain and the interval in which part of the remaining space was filled with blocky cement. The *time of diagenesis* is thus defined on a relative time scale. Close inspection may allow a greater refinement of the time interval of diagenesis. For instance, grains may be attacked by algae, which often show up as a dark, fuzzy rim around the grain "under" the drusy, acicular cement crystals. This subdivides the period before the first cementation into a time without and a time with algal activity. Iron coatings around certain blocky cement crystals and/or the presence of iron in the lattices of certain blocky cement crystals (as revealed by staining) could alternate with a total absence of iron. After precipitation of the drusy acicular cement crystals, this could suggest periods during which iron could be taken up in the crystal lattices and periods in which this was not the case.

The hypothetical example in Fig. 3–1 demonstrates the possibility of correlating a certain moment in the sequence of diagenetic events with a certain moment in the palaeogeographic development. For instance, if an iron-rich hinterland is activated, then it is reasonable to relate the sudden presence of iron in the fabric of the rock to such a tectonic event. Thus by careful study, the timing of diagenetic and of structural geological events might be related.

Zone in Which Particular Diagenetic Processes Act

The field of diagenesis is bound by three definite junctions and forms part of the burial environment. Within the field of diagenesis a multitude of processes act, depending on the nature of the pore-filling fluid, the physical parameters related to depth and the mineralogical composition of the rock. Amongst the pore-filling fluids, meteoric, marine and hydrothermal water and hydrocarbons are the most significant ones. Mixtures of various fluids result in variations in reactivity of such mixtures and thus in various effects on the solubility of certain minerals (see p. 234). The depth setting for diagenetic processes expresses itself in zones in which air, or moving or stationary fluid occurs. The latter zone type can be subdivided into intervals in which pressure and temperature effects are discernible and those in which this is not the case. Thus the field of diagenesis can be subdivided into ten diagenetic environments occurring in three zones, as shown in Fig. 3–2.

Effect of Time on the Nature of Diagenetic Processes: Stages in Diagenesis

Overprinting of one another by diagenetic features allows, on a relative time scale, the establishment of diagenetic sequences. In specific cases it might allow correlation of diagenetic with other geological events on a relative or absolute time scale. But juxtaposition of diagenetic products also reveals the diagenetic environments in which the processes took place. Variations in intensity of diagenetic imprints may be noticed in diagenetically altered sediments. Such degrees of modification of the sediment could give an idea of the duration of the diagenetic processes that modified the sediments or alternatively of the susceptibility of the sediments to diagenetic processes. The establishment of the duration of diagenetic processes or of the sediment susceptibility is thus mainly deductive, and detailed microscopic investigation is required. The aim of such investigations is recognition of the stage of alteration and further delineation of the diagenetic environment. The recognition of stages in diagenesis is meaningful with parameters that reflect a range of intensities and leave easily detectable diagenetic products. One such parameter is the overburden pressure which characteristically modifies sediments. It is a major criterion for the recognition of diagenetic environments in which the following stages of alteration are recognised "in time":

1. Syndepositional stage: The period during which sediments undergo diagenetic processes that are directly controlled by conditions in the depositional environment (see p. 181).
2. Synerosional stage: The period during which sediments (with or without a diagenetic history) are subjected to diagenetic processes (weathering and soil formation) that are directly controlled by conditions in the erosional environments (see pp. 181–185). Modifications of the sediment that take place "in time" during the syndepositional or synerosional stage occur "in space" within the *surface-related diagenetic zone*.

3. Syncementation stage: The period during which sediments are subjected to diagenetic processes that act concurrently with cement formation (pp. 181–185). Depending on the nature of the cement formed, such processes can be pinpointed within the surface-related diagenetic zone or within the *shallow burial* diagenetic environment.
4. Post–pressure solution stage: The period during which sediments are subjected to diagenetic processes that post-date pressure solution features, notably stylolite formation (pp. 185–186).

In space such processes occur in the *deep burial* diagenetic environment. It is possible, and indeed rather common, to find sediments with signs of two or more diagenetic stages. This is possible because with ongoing overburden accumulation (or removal) such a sediment "travels" through various diagenetic environments. Moreover, it might have various susceptibilities to different diagenetic processes.

The establishment of diagenetic stages is not an aim in itself. It helps to determine which processes take place in which diagenetic environment. Processes leading to increased porosity or permeability can be predicted more accurately if their relation to the environments in which they act is understood.

Conclusions

The field of diagenesis (Fig. 3–2) is characterised by three aspects of diagenesis: (1) the time interval during which diagenetic processes act; (2) the zone in which particular diagenetic processes act; and (3) the time effect provided by the sequential nature of the diagenetic processes, also known as *stages in diagenesis*.

The accurate recognition of diagenetic environments or zones in ancient rocks helps one to understand the porosity-creating or porosity-destroying diagenetic processes (Fig. 3–3) and the presence or absence of reservoirs. Diagenetic environment recognition in ancient rocks is facilitated by comparative studies in Recent sediments in which similar diagenetic fabrics, structures or minerals occur which can be linked with Recent diagenetic and depositional environments. This connection forms the basis for extrapolation. Diagenetic features (fabrics, minerals, structures) may be characteristic of certain depositional facies when these are more susceptible to particular diagenetic processes because permeability is high. Then, associations of diagenetic and depositional features result in *carbonate fabric units* (Reijers and Bartok, 1985). The definition of carbonate fabric units (Definition 300) and their regional recognition allows the regional tracing of potential reservoirs.

In the near-surface diagenetic zone, processes are triggered by parameters in the depositional environment, giving rise to lateral variations in diagenetic products. In this diagenetic zone a distinction can be made between terrestrial, marine or mixed-zone diagenetic environments in which vadose and phreatic diagenetic products allow a further zonal refinement. In a downward direction the occurrence of pressure solution phenomena is related to increasing overburden. Thus the first occurrence of stylolites marks the end of the shallow burial diagenetic environment, although other pressure solution features occur at shallower levels.

A correct determination of the overprinted sequence of diagenetic products helps to establish the diagenetic environments through which sediments have passed. Establishing the sequence of

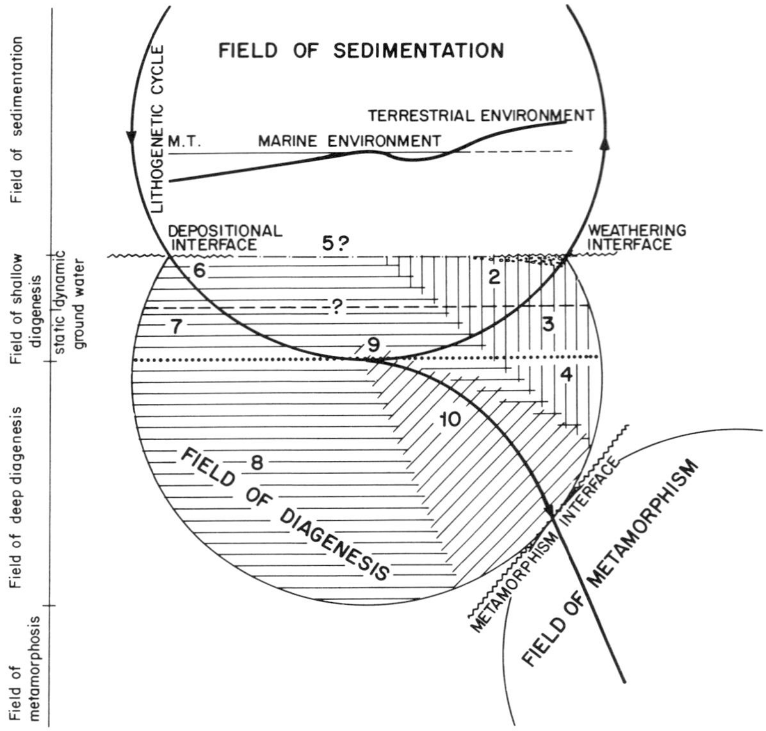

NATURE OF PORE-FILLING FLUID

dynamics of pore-filling agent — increasing depth	METEORIC	MARINE	HYDROTHERMINAL
AIR	vadose diagenetic environments 1	5	
DYNAMIC WATER	terrestrial dynamic diagenetic environment 2	marine dynamic d.e. 6	hydrothermal dynamic d.e. ?
STATIC WATER	terrestrial static d.e. 3	marine static d.e. 7	hydrothermal static d.e. 9
STATIC WATER (DEPTH INFLUENCES)	terrestrial deep static d.e. 4	marine deep static d.e. 8	hydrothermal deep static d.e. 10

Diagenetic environments within the field of diagenesis. The environments are numbered, and these numbers correspond with those in the table.

– – – Transition boundary between the zone of static and the zone of dynamic ground water in the interval of shallow diagenesis.

······· Transition zone between the interval with static ground water with and without influence of depth factors. Moreover, below this zone one notices the increased importance of hydrothermal water

Fig. 3–2. Diagenetic environments (d.e.) within the field of diagenesis.

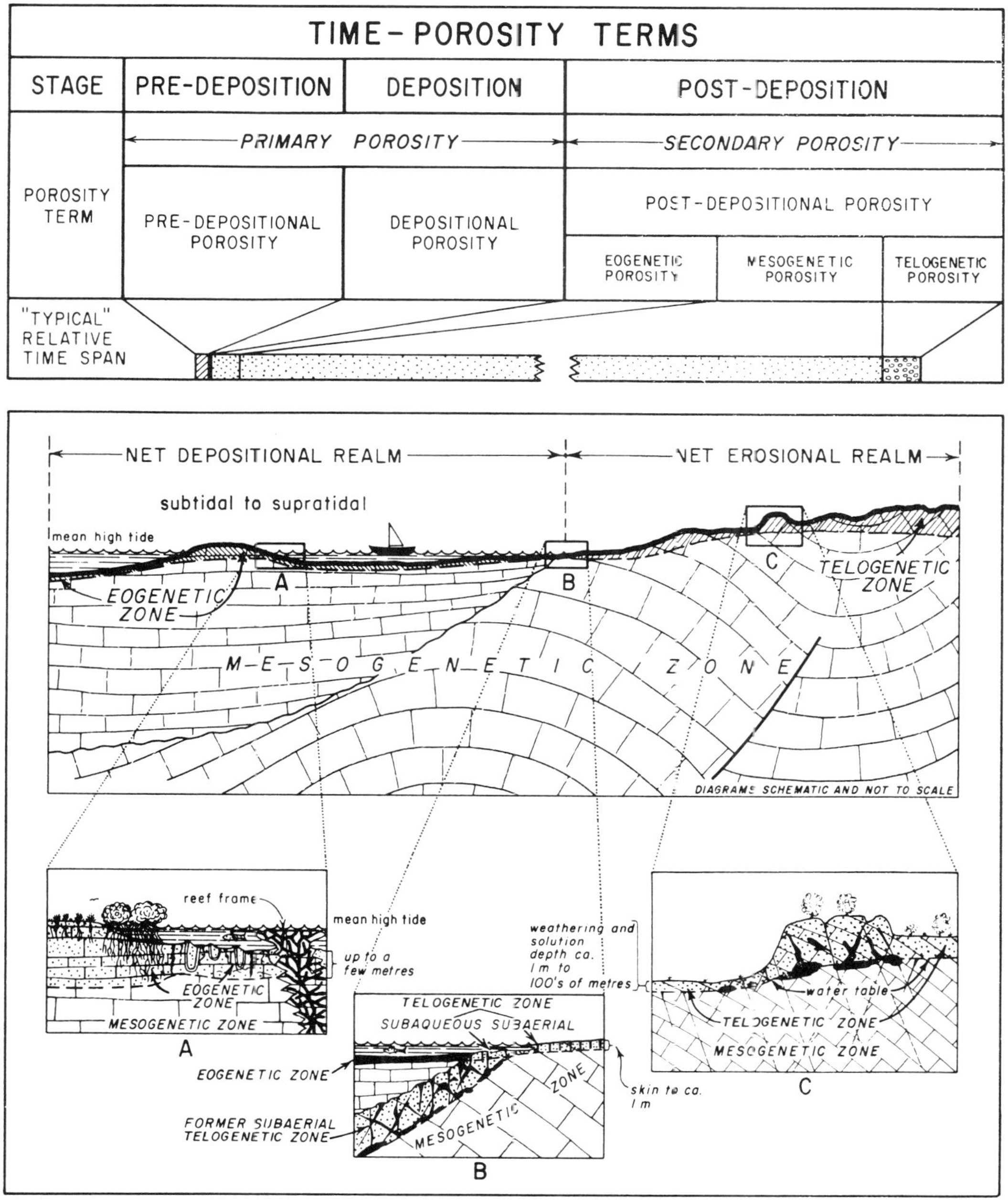

Fig. 3–3. Time–Porosity terms and zones of porosity modification in sedimentary carbonates. Illustration by permission of AAPG.

activity of diagenetic processes is often not very difficult; deducing the actual time interval during which these processes acted is more difficult. If such a deduction is the aim, then knowledge of other (regional) geological events is pertinent. Such events should be tied into the petrographical data. Establishing the accurate time interval during which porosity-enhancing and porosity-occluding processes were active is important, because this determines the time during which the "reservoir window" occurred. Structural studies help to establish the period in which structural traps were formed, and geochemical studies help to establish the period during which source rocks could generate hydrocarbons. Such studies should be complemented by an analysis of the time interval over which certain diagenetic processes were active; only then can a full prospect evaluation be carried out. Such studies also help to define stratigraphical traps of a diagenetic nature which could occur as a result of porosity-creating and porosity-occluding processes being active simultaneously in different areas. Commonly, *diagenetic time* is expressed in the overprint of diagenetic marks in the rocks. Younger marks overprint older ones. Thus diagenetic studies should be geared towards recognising the diagenetic sequence that can be deduced from the diagenetic paragenesis. The subdivision into early, advanced and late diagenetic processes or products is often parallel to that of near-surface, shallow and deep diagenesis, but it is not necessary that this always be so. Time is recorded as overburden only in the case of uninterrupted, constant sedimentation. In the case of condensed sequences, exceptionally fast infills and/or erosion, which are all very common in the geological record, it is not possible to parallelise time and depth.

Diagenetic processes are variables within diagenetic environments. Their intensity varies in time and space. In this respect they are directly comparable to depositional processes in depositional environments. The marks left by diagenetic processes can be used to recognise space and time in diagenesis. Variations in activity of diagenetic processes may give rise to differential diagenesis. Such variations could have a time or a space aspect. Variations in the nature of diagenetic processes give rise to destructive versus constructive diagenesis. Variations in the kind of diagenetic processes allows their classification as chemical, physical, biological, biochemical, mechanical, etc.

Considerations of the various aspects of diagenesis are not an aim in themselves but should be regarded as a step in the process that leads ultimately to a fuller understanding of what happened, why and when.

Definitions 269–300 (General Diagenetic Concepts; see Fig. 3–4)

269. DIAGENESIS (PROCESSES): all chemical, physical and biogenic processes that take place in a sediment after its deposition and before metamorphism.

NE diagenese
DE Diagenese
ES diagénesis
FR diagénèse

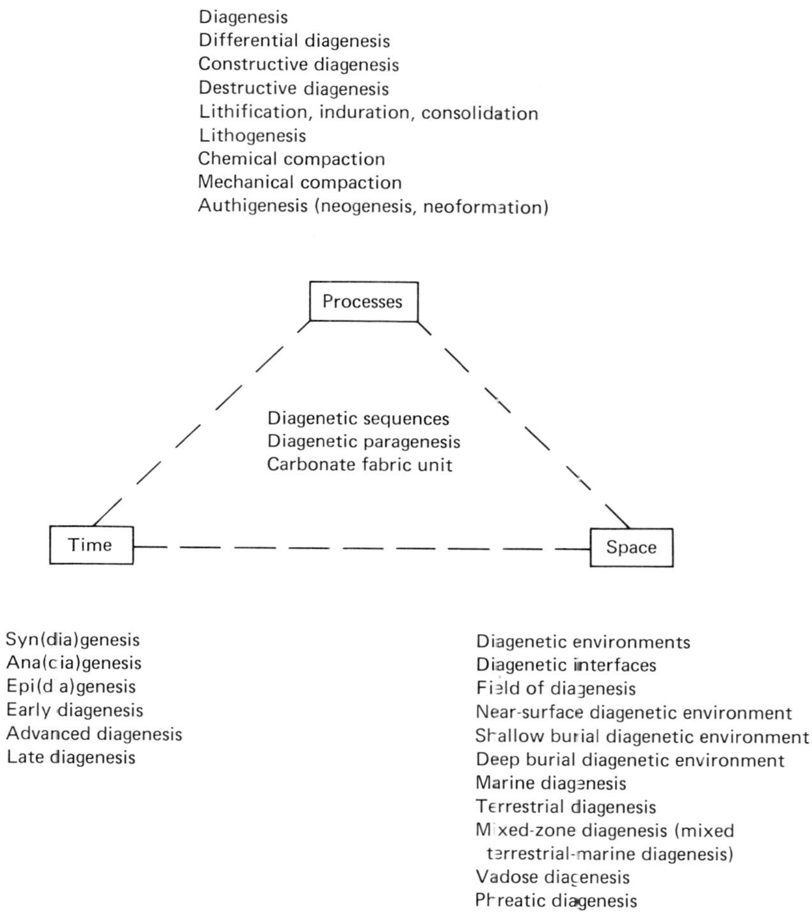

Fig. 3–4. Groups of general diagenetic terms.

270. DIAGENETIC: related to diagenesis.

NE diagenetisch
DE diagenetisch
ES diagenética
FR diagénétique

271. DIFFERENTIAL DIAGENESIS: a difference in diagenetic history, expressed by the presence of different diagenetic paragenesis in depositionally identical sediments. An example of differential diagenesis is the introduction of liquid

"live" hydrocarbons into a reservoir which inhibits cementation in the hydrocarbon-bearing zone but does not inhibit cementation in the water-bearing zone, and within intervals containing "dead" hydrocarbons.

NE differentiële diagenese
DE differentielle Diagenese
ES diagénesis diferencial
FR diagénèse différentielle

272. CONSTRUCTIVE/DESTRUCTIVE DIAGENESIS: diagenesis due to the interaction between water and sediment, respectively resulting in precipitation/cementation versus dissolution/breakdown of material. The type of diagenesis indicates the direction of the diagenetic flux of carbonate matter. In the constructive type of marine diagenesis, seawater acts as the donor from which carbonates are extracted; in the destructive type, sediments act as the donor from which carbonates are extracted while the ambient seawater is the recipient.

Note: See Alexandersson (1973, 1974). For the precise meaning of *dissolution*, see Definition 301.

NE opbouwende/afbrekende diagenese
DE konstruktive/destruktive Diagenese
ES diagénesis constructiva/destructiva
FR diagénèse constructrice/destructrice

273. LITHOGENESIS: the origin and formation of indurated, consolidated rocks, especially sedimentary rocks; also the science of the formation of rocks.

Note
1. Lithogenesis encompasses lithification (Definition 274), induration (Definition 275) and consolidation. Important processes leading to lithification, induration or consolidation are chemical and/or mechanical compaction (Definition 276) and autigenesis, neogenesis and neoformation (Definition 298).
2. See also Definition 405.

NE lithogenese
DE Lithogenese
ES litogénesis
FR lithogénèse

274. LITHIFICATION: see induration.

Note: See note to Definition 273.

NE vastwording; verharding; verstening
DE Lithifikation
ES litificación; induración
FR lithification; induration

275. INDURATION (CONSOLIDATION; LITHIFICATION): a general term referring to the diagenetic processes whereby sedimentary rocks are consolidated from a state of loose aggregation.

Note: See note to Definition 273.

NE vastwording; verharding; lithifikatie; steenwording
DE Verfestigung
ES litificación; induración; consolidación
FR lithification; induration; consolidation

276. COMPACTION: volume reduction of a sediment or rock accompanied by a loss of pore volume and/or a loss of rock mass chemical and physical processes give rise to chemical and mechanical compaction.

Note: See note to Definition 273; see also Definitions 315 and 316.

NE compactie
DE Kompaktion
ES compactación
FR compaction

277. SYNDIAGENESIS (SYNGENESIS): diagenetic processes occurring at approximately the same time as sedimentation.

Note: Syndiagenetic is also used to indicate the contemporaneous occurrence of diagenetic processes at any stage of diagenesis, but this usage is not recommended.

NE syndiagenese; syngenese (vroege diagenese)
DE Syndiagenese
ES sindiagénesis
FR syngénèse

278. ANADIAGENESIS (ANAGENESIS): diagenetic processes occurring during the phase of diagenesis coinciding with compaction or increase in the overburden of the sediment.

NE anadiagenese; anagenese (intermediaire diagenese)
DE Anagenese
ES anadiagénesis
FR anagénèse

279. EPIDIAGENESIS (EPIGENESIS): diagenetic processes affecting sedimentary rocks subsequent to their compaction and/or pressure solution, exclusive of weathering and metamorphosis; it thus coincides with late diagenesis.

NE epidiagenese; epigenese–late diagenese
DE Epigenese
ES epigénesis; epidiagénesis
FR épigénèse

280. SYNDIAGENETIC (SYNGENETIC): early diagenetic.

Note: See note to Definition 277.

NE vroeg diagenetisch
DE syngenetisch
ES singenético; sindiagenético
FR syngénétique

281. ANADIAGENETIC (ANAGENETIC): intermediate diagenetic.

NE intermediair diagenetisch
DE anagenetisch
ES anagenético; anadiagenético
FR anagénétique

282. EPIDIAGENETIC (EPIGENETIC): late diagenetic.

NE laat diagenetisch
DE epigenetisch
ES epigenético; epidiagenético
FR épigénétique

283. FIELD OF DIAGENESIS: that part of the realm of lithogenesis which is bounded by the depositional, weathering and metamorphism interfaces (see Fig. 3–2). It can be subdivided into the near-surface, shallow burial and deep burial diagenetic environments.

NE diagenetisch interval
DE Diagenesebereich
ES el campo de la diagenesis; dominio diagenético
FR domaine diagénétique; domaine de la diagénèse

284. FIELD OF SEDIMENTATION: that part of the realm of lithogenesis in which transportation and deposition of sedimentary particles occur.

NE sedimentatie gebied
DE Sedimentationsbereich
ES el campo de la sedimentación; dominio sedimentario
FR domaine sédimentaire; domaine de la sédimentation

285. FIELD OF EROSION: that part of the realm of lithogenesis in which sedimentary particles or rocks are formed or modified under the influence of atmospheric or subaquatic environmental conditions.

NE erosie gebied
DE Erosionsbereich
ES el campo de la erosión; dominio erosivo
FR domaine de l'érosion

286. FIELD OF METAMORPHISM: that part of the realm of lithogenesis in which physicochemical alterations of the rock obliterate sedimentary structures and in which distinctive new associations of minerals occur, alien to the field of sedimentation.

NE metamorphose gebied
DE Metamorphosebereich
ES el campo del metamorfismo; dominio metamórfico
FR domaine métamorphique; domaine du métamorphisme

287. DEPOSITIONAL INTERFACE: the junction between the fields of sedimentation and diagenesis.

Note: Depositional interface could be defined in a more limited sense as the interface between the water and the bottom where sediments are deposited (Plumbley *et al.*, 1962). In that sense the German term *Ablagerungsfläche* is preferred.

NE afzettingsoppervlakte
DE Ablagerungsfläche; Grenzbereich, übergangsbereich

ES superficie de sedimentación
FR interface sédimentation–diagénèse

288. WEATHERING INTERFACE: the junction between the fields of erosion and diagenesis.

NE verweringsoppervlakte
DE Verwitterungsfläche
ES superficie de meteorización
FR interface altération–diagénèse

289. METAMORPHISM INTERFACE: the junction between the fields of metamorphism and diagenesis.

NE metamorfose oppervlakte
DE Metamorphosefläche
ES superficie de metamorfismo
FR interface diagénèse–métamorphisme

290. DIAGENETIC ENVIRONMENT: a part of the field of diagenesis that is typified by assemblages of characteristic diagenetic features (see Figs. 3–5, 7, 8).

Note: A specific diagenetic environment is characterised by specific products of diagenetic processes that act in that environment. With ongoing lithogenesis, the rock could be subject to other diagenetic processes, acting, for example, later or deeper and reflecting another diagenetic overprint.

NE diagenetische omgeving
DE diagenetisches Milieu
ES ambiente diagenético
FR milieu diagénétique

291. DIAGENETIC SEQUENCE: the sequential order of formation of diagenetic products (see Fig. 3–6).

NE diagenetische volgorde
DE diagenetische Folge; diagenetische Sequenz
ES secuencía diagenética
FR séquence diagénétique

292. DIAGENETIC PARAGENESIS: a characteristic association of diagenetic products (see Fig. 3–7).

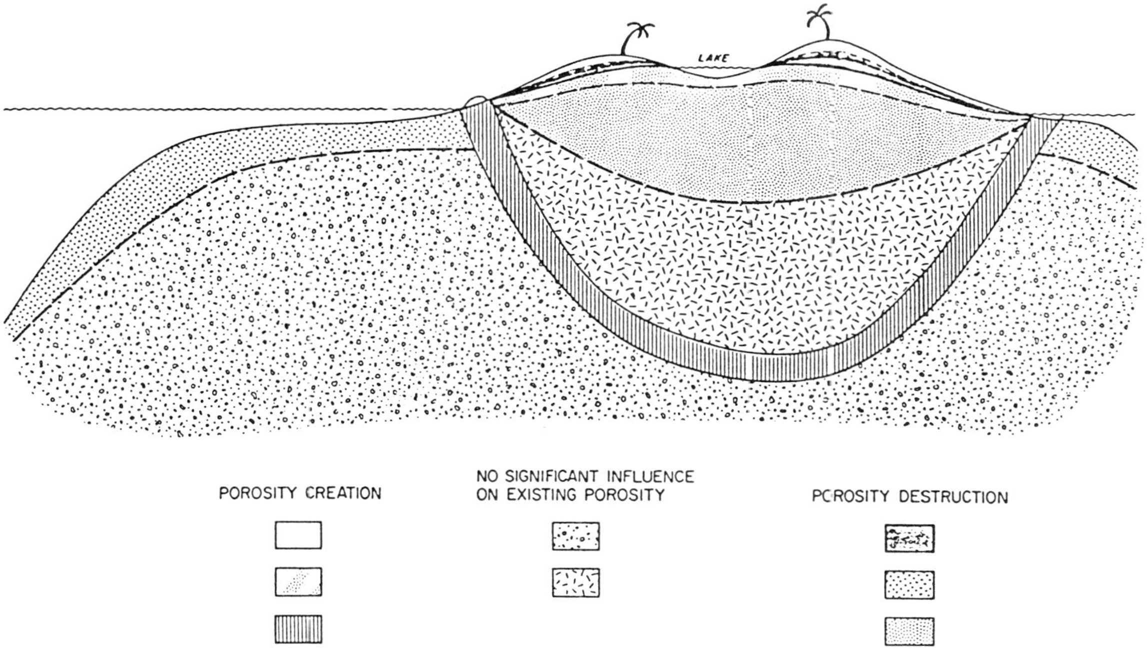

Fig. 3–5. Diagrammatic representation of the environments of diagenesis: the effect on porosity modification.

Note
1. Ideally a diagenetic paragenesis reflects a specific diagenetic environment, but see note to Definition 290.
2. A diagenetic paragenesis might also reflect an ongoing diagenetic process which leaves in time and space different diagenetic products (e.g., pressure solution, which might successively result in grain interpenetration, horsetails and stylolites).

NE diagenetisch gezelschap
DE diagenetische Paragenese
ES paragénesis diagenética
FR paragenèse diagénétique

293. NEAR-SURFACE DIAGENETIC ENVIRONMENT: a diagenetic environment in which processes act that are related to the depositional or weathering interface (see Fig. 3–2).

Note: See Harrison (1977).

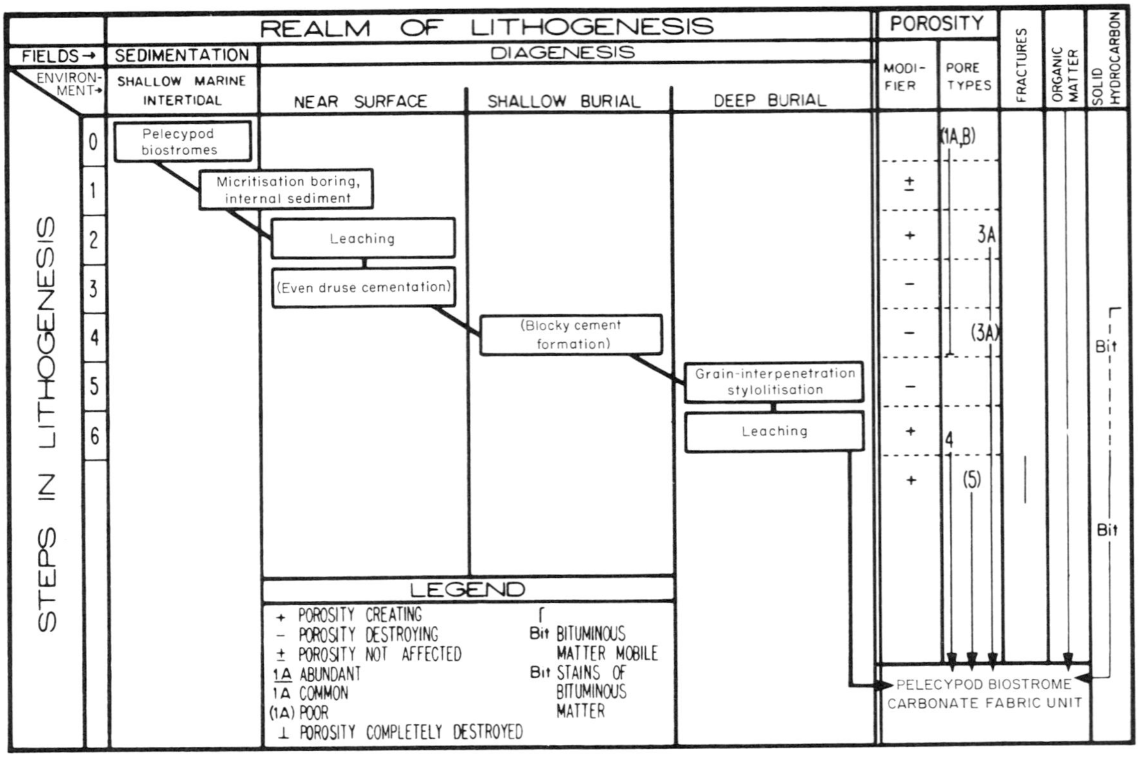

Fig. 3–6. Diagenetic model in which a diagenetic sequence of processes is deduced from a paragenesis of products. These have been placed in time and space order.

NE oppervlakte-nabije diagenetische omgeving
DE oberflächennahes Diagenese-Milieu
ES ambiente diagenético cerca de la superficie
FR milieu diagénétique superficiel

294. SHALLOW BURIAL DIAGENETIC ENVIRONMENT: a diagenetic environment in which processes act that are not entirely related to the depositional environment and that predate stylolite formation.

NE ondiep begraven diagenetische omgeving
DE diagenetisches Milieu mit geringmächtiger Sedimentbedeckung
ES ambiente diagenético de poca profundidad
FR milieu de la diagénése précoce; milieu diagénétique superficiel; peu profond

FILL	COMPOSITION	DYNAMICS	SATURATION	PROCESSES	PRODUCTS
VADOSE	MARINE			EXTENSIVE ARAG. / CALC. CEMENTATION	BEACH ROCK (MENISCUS / PENDANT CMT.) FIBROUS SHAPE
VADOSE	FRESH WATER		UNDERSATURATED	1. SOLUTION BY UNDERSATURATED METEORIC WATER 2. PRODUCTION OF CO_2 IN SOIL ZONE AIDING SOLUTION	PREFERENTIAL REMOVAL OF ARAGONITE RESULTS IN VUGS
VADOSE	FRESH WATER		SATURATED	1. MENISCUS / PENDANT DISTRIBUTION OF WATER 2. CO_2 LOSS (EVAPORATION)	CEMENTS {MENISCUS, PENDANT} EQUANT CALCITE
PHREATIC	FRESH WATER	ACTIVE	UNDERSATURATED	SOLUTION BY UNDERSATURATED METEORIC WATER	1. MOULDIC AND/OR VUGGY POROSITY 2. NEOMORPHISM (INVERSION ARAGONITE → CALCITE) IN GRAINS WITH UNSTABLE MINEROLOGY
PHREATIC	FRESH WATER	ACTIVE MOD. ACT.	SATURATED SATURATED	1. RAPID TO MODERATE WATER CIRCULATION 2. LIMITED LEACHING (OF ARAGONITE)	1. CALCITE CEMENTS {EQUANT, SYNTAXIAL, ISOPACHOUS BLADED} 2. ARAGONITE → CALCITE INVERSION 3. INTERLOCKING CEMENT CRYSTALS 4. LOW POROSITY
PHREATIC	FRESH WATER	STAGNANT	SATURATED	1. REDUCED OR NO WATER CIRCULATION	1. LITTLE CEMENT / LEACHING 2. PRESERVATION OF PREEXISTING ∅ 3. STABILISATION OF Mg - CALCITE AND ARAGONITE
PHREATIC	BRACKISH (MIXED)	RELATIVELY STAGNANT, BUT SOME MIXING	SAT. - UNSAT. WITH RESPECT TO $CaCO_3$ OVERSAT. WITH RESPECT TO $CaMg(CO_3)_2$	1. DOLOMITISATION	1. DOLOMITE 2. PRODUCTS AS IN FRESH / MARINE ZONES
PHREATIC	MARINE	ACTIVE	USUALLY SATURATED	1. WATER CIRCULATION THROUGH SEDIMENTS	ARAGONITE {NEEDLES, FIBROUS ISOPACHOUS (POLYGONAL BOUNDARIES), BOTRYOIDAL} Mg - CALCITE {MICRITIC, FIBROUS ISOPACHOUS (POLYGONAL BOUNDARIES), PSEUDOPELLETS} CEMENT - SEDIMENT INTERBEDS; CEMENTS BORED
PHREATIC	MARINE	STAGNANT	SATURATED	1. REDUCED OR NO WATER CIRCULATION 2. BACTERIAL (?) CONTROL ON CEMENTATION	1. MICRITISATION 2. LITTLE CEMENTATION EXEPT IN SKELETAL MICROPORES

Fig. 3–7. Some diagenetic processes and products in idealised environments of diagenesis.

295. DEEP BURIAL DIAGENETIC ENVIRONMENT: a diagenetic environment in which processes act that post-date stylolitisation. This feature therefore characterises its upper limit; the metamorphism interface is its lower limit.

NE diep begraven diagenetische omgeving
DE diagenetisches Milieu mit mächtiger Sedimentbedeckung
ES ambiente diagenético profundo
FR milieu diagénétique profond

296. VADOSE DIAGENESIS: all post-depositional changes in sediments the pore space of which is partly water and partly air filled. Such changes usually occur around the water table in a sediment near the depositional and weathering interface.

NE vadose diagenese
DE vadose Diagenese
ES diagénesis del tipo vadoso
FR diagénèse vadose

297. PHREATIC DIAGENESIS: all post-depositional changes in sediments the pore space of which is entirely water filled.

NE freatische diagenese
DE phreatische Diagenese
ES diagénesis freática
FR diagénèse phréatique

298. AUTHIGENESIS (NEOGENESIS; NEOFORMATION): formation of minerals in place within a sediment or sedimentary rock.

Note: See note to Definition 273.

NE autigenese; nieuwvorming
DE Authigenese
ES autigénesis
FR authigénèse; néogénèse; néoformation

299. AUTHIGENIC (AUTHIGENETIC; AUTHIGENOUS; NEOFORMATED): said of constituents (especially mineral constituents) of sedimentary rocks that are formed in place.

NE authigeen
DE authigen

ES autigénico; autígeno
FR authigénique; authigène; néoformé

300. CARBONATE FABRIC UNIT: a carbonate rock interval characterised by uniform depositional conditions and a characteristic diagenetic history. It is thus a homogeneous facies unit, rock unit or lithotype, comprising characteristic depositional and diagenetic features and structures.

NE karbonaat maaksel eenheid
DE Faziell und diagenetisch einheitliche Karbonateinheit
ES unidad textoral carbonatada
FR unité textural

3–3 DIAGENETIC PROCESSES

Introduction

The motion of subsurface water is governed by the laws of physics; the chemical composition by the laws of chemistry. Moving subsurface water flows continuously into a new chemical environment, where, by reaction with its enclosing medium, a chemical equilibrium is achieved. Such reactions include dissolution, leaching (Definition 301) and karstic solution, but also precipitation (cementation) or simultaneous solution and precipitation (replacement). Diagenetic changes are thus a natural consequence of subsurface hydrodynamic movement. They ought to occur, sooner or later.

Early diagenesis occurs shortly after the deposition of a sediment, usually in a surface-related or shallow burial diagenetic environment. The relief, climate and other factors influencing the depositional environment (Fig. 2–2) are similar to those influencing the diagenetic environment. Thus, early diagenetic products are frequently distributed along depositional facies trends.

Subsequent diagenesis occurs (long) after the deposition of a sediment, frequently under overburden conditions and occasionally in surface-related environments (weathering).

The factors influencing the nature and movement of the diagenetic solution in deep waters differ considerably from those that influence sedimentation. Thus, deep diagenetic products are indirectly and remotely related, if at all, to the distribution of the depositional facies and only so if permeability remains in such facies.

Early Diagenetic Processes

Calcium carbonate sediment (see, for mineralogy, Definitions 3 and 4) can be altered by dissolution, leaching, precipitation, recrystallisation and replacement. A loose carbonate sediment changes to a consolidated limestone by cementation. Waters precipitating cement at one time might under other circumstances cause dissolution or recrystallisation. Thus dissolution, precipitation and recrystallisation of a limestone are often closely related expressions of flow of ground waters.

A calcium carbonate sediment might be affected by a mineral brine under near-surface conditions, being thus changed to another mineral. Dolomitisation is the most common of such processes. The nature of dolomitising water and its movement paths have been the subject of much discussion.

Solution/Precipitation. Calcium carbonate can be either dissolved or precipitated in the vadose and phreatic zones of subsurface water circulation. Diagenetic processes include solution, dissolution, leaching, chemical compaction, fracturing and brecciation, internal sedimentation, laminar and concretionary growth and cementation. The diagenetic products include inter- and intraparticle pores, caves, vadose compaction features, crumbly fractures, solution collapse breccia, vadose silt, caliche, pisoid, dripstone, travertine, drusy, blocky and rim cement, etc. (see also p. 234 and Fig. 3–8).

Solution, dissolution, leaching and precipitation of calcium carbonate take place mainly in the shallow burial and/or surface-related diagenetic environments along the paths of ground-water flows. The distribution of the various diagenetic products is not straightforward since both destructive and constructive diagenetic processes are active. One must determine in which places solution, leaching and dissolution predominate, creating and enlarging pore space, and in which places cementation predominates, occluding pore space.

From core investigations it has repeatedly been demonstrated that in the shallow phreatic zone extensive dissolution occurs. This is consistent with experimental and theoretical deductions. Solubility relations suggest downward percolating waters dissolving carbonate and calcium carbonate until, in the realm of phreatic flow, the water becomes saturated. Studies on the rate of solution suggest that ground water entering limestones quickly becomes saturated with calcium carbonate in the zone of phreatic flow. Any additional carbonate can be taken on by this water only if its capacity to dissolve carbonate increases. This (and associated leaching) occurs only if the water flows in the direction of increasing solubility (i.e., upwards). Descending waters of constant salinity move in the direction of decreasing solubility. Precipitation rather than solution by descending waters should be the rule in the deep phreatic zone. Thus the vadose and shallow phreatic zones are favoured as the zones of extensive solution. Here, the subsurface waters are not yet saturated with calcium carbonate. In the deeper part of the vadose and phreatic zone, however, various other processes may result in the elimination of pore space.

An understanding of these processes aid in avoiding cemented carbonate reservoirs and locating regions in which cementation is minor or absent. Calcite cement in the vadose and phreatic zones has been explained by the following:

1. Evaporation and escape of carbon dioxide in the vadose zone. This could result in precipitation of calcium carbonate.
2. Descending water in the phreatic zone, flowing in the direction of decreasing solubility. This also triggers precipitation of calcite.

Aragonite and calcite cements in the vadose zone often tend to be finely crystalline or drusy. Calcite cement in the phreatic zone tends to be medium to coarsely crystalline. The pore space in carbonates is usually entirely occluded after precipitation of a first generation of drusy cement and infill by early internal sediment followed by a second generation of blocky cement precipitation. But when conditions are unfavourable for the formation of the second generation of phreatic cement, pore space can be preserved. The formation of phreatic cement involves transfer of material; consequently a cessation of water movement reduces the efficiency of such transfer. Under hydrostatic conditions, transfer of material in a solution takes place only along a concentration gradient by diffusion. Thus phreatic cementation becomes ineffective when a carbonate is moved out of the zone of phreatic circulation. This may happen after uplift or submergence. A carbonate bed can be uplifted into the vadose zone, where the pore space in the carbonate commonly is partly air filled. Under the right condition this may remain so. At present only a limited number of data are available on the distribution of phreatic cements in ancient carbonates. Mapping of distribution trends, perhaps after storage of the data in a computer-compatible way (see Section 4–4) and statistical manipulation and cross-retrieval of the data base could in the near future provide the critical evidence needed to understand cementation distribution patterns fully.

Replacement. Dolomitisation is the most common replacement process, and it can be significant for the formation of reservoirs. The origin of dolomite has been a subject of considerable controversy in geological literature. Almost all dolomite is of replacement origin, but the processes responsible for dolomitisation are a topic of speculation. Understanding those processes, however, might improve predictions of the occurrence of dolomite, in particular porous dolomite.

DISTRIBUTION PATTERNS OF DOLOMITE. Ancient dolomites occur frequently in the following regional patterns:

1. In the uppermost beds of the shelf facies of a carbonate sequence. Being situated on top of a carbonate platform increases the chance of a carbonate being dolomitised, but this condition alone is not sufficient.
2. In association with gypsum or anhydrite deposits. The conditions favourable for the deposition of gypsum are also favourable for the formation of dolomite, but such conditions do not always lead to the formation of dolomite.
3. Beneath regional unconformities. Subaerial exposure itself, however, does not constitute a sufficient condition for the formation of dolomite.

Selective dolomitisation in ancient dolomites occurs frequently in the following local patterns:

1. Certain permeable biofacies
2. More permeable zones (burrows, fractures, faults)
3. Muddy lime deposits

(Sub)recent dolomites have been discovered in the following:

1. Saline playa deposits (uncertain whether replacement or primary precipitate)
2. Supratidal zones (replacement in origin)
3. Lagoons (primary precipitate?)

Although Recent dolomites have been found in several different settings and environments and their origin has been variously interpreted, the occurrences share the following characteristics:

1. No dolomite occurs in normal marine depositional environments. Almost all dolomites occur in an intertidal setting or environment.
2. Almost all occur in arid or semiarid areas.

3. In places, dolomite is associated with gypsum; elsewhere, with other magnesium minerals (e.g., hydromagnesite). In a few places, dolomite is apparently not associated with gypsum or other evaporite minerals.

CHEMICAL CONSIDERATIONS OF DOLOMITISATION. A solution must have a certain magnesium/calcium concentration ratio before it replaces calcite and forms dolomite. The ratio is not constant under a given temperature and pressure because the presence of various other ions in a solution could change its value. It is the value of the ratio that defines the conditions for dolomitisation. These conditions can be generalized as follows:

1. The higher the magnesium/calcium concentration ratio of a fluid, the more likely it is that it will become dolomitising.
2. The low magnesium/calcium concentration ratios of most ground waters apparently in equilibrium with both calcite and dolomite suggest that the concentration ratio required for ground water to dolomitise is not high. It is perhaps not greater than 1.
3. The high magnesium/calcium concentration ratio of seawater (± 5.2) and the apparent inability of normal marine seawater to replace calcium carbonate with dolomite suggest that either (1) dolomite is formed by the seawater at a very low rate or (2) the presence of various ions in seawater necessitates a very high magnesium/calcium concentration ratio for seawater to dolomitise.
4. The end result of evaporation of seawater is a solution very rich in magnesium from which various salts will form.

WORKING HYPOTHESES FOR DOLOMITISATION PROCESSES. Dolomitisation involves the transfer of materials. Any working hypothesis of the origin of secondary dolomite must postulate (1) a dolomitising solution capable of reacting with calcite to form dolomite and (2) a mechanism which enables the dolomitising solution to move through the sediment. Three different working hypotheses, and the probable implications of each, can now be formulated with these postulates. They explain dolomitisation in ancient carbonates as the result of an intrastratal solution flowing through a porous medium. These hypotheses, each of which has merits and shortcomings, are as follows:

1. Dolomitisation by Seepage Reflux: Halite is scarce in thick gypsum or anhydrite deposits, suggesting that highly concentrated brines escaped before complete desiccation. It is generally assumed that heavy brines return to the ocean by flowing along the bottom of lagoons (surface reflux) or through porous and permeable sediments underlying the lagoon (seepage reflux). Such brines have a high magnesium/calcium concentration ratio. Consequently they should be able to react with calcite to form dolomite.

2. Dolomitisation by Evaporative Pumping: Recent and ancient dolomites have been found in the realm of subaerial exposure where capillary vadose and/or phreatic waters are present. These occurrences suggest that such subsurface waters could, under certain circumstances, react with brine sediment to form dolomite.

Phreatic waters are commonly very poor in magnesium and usually have magnesium/calcium concentration ratios below 1. Evaporation brings water to the surface, where brines are formed, and the magnesium/calcium ratio increases considerably. Any such subsurface water is unstable if the region contains calcite. This is so because the water establishes equilibrium with the calcite, by dissolving calcite, by directly precipitating dolomite or by replacing calcite with dolomite. Subaerially exposed carbonates (commonly the top beds of a carbonate platform) tend to be preferentially dolomitised, because vadose and phreatic water movements due to capillary action are restricted to land areas. The continuously submerged carbonates forming the slope of carbonate platforms escape this type of dolomitisation, because their interstitial fluids are essentially hydrostatic.

3. Dolomitisation by Mixing Seawater and Fresh Water: Dolomitisation by ground water is hampered by a low replacement rate and a low magnesium/calcium concentration ratio. However, when fresh phreatic (meteoric) water is mixed with up to 30% seawater, the mixture may become undersaturated with respect to calcite. If 5–30% seawater is admixed, the saturation with respect to dolomite may be considerable and dolomitisation by replacement could occur. Vertical changes in the position of the brackish water zone are the result of sea level fluctuation and changes in supply of fresh water. These may result in considerable volumes of rock being dolomitised around the (fluctuating) water table.

It is not always clear what the source of magnesium is for dolomitising fluids. If vadose and phreatic waters can react with brine to form dolomite, it is still uncertain whether the carbonate sediment is permeable enough to permit a sufficient flow of these waters to dolomitise large volumes of lime deposits.

Subsequent Diagenetic Processes

Early diagenetic processes result in early lithified carbonate sediments due to the flow of subsurface water. Later diagenetic processes also occur when subsurface water reacts with its enclosing medium during the course of its movements. Consequently later diagenetic processes also include solution, precipitation, recrystallisation and replacement.

Solution/Precipitation of Calcium Carbonate. Factors governing the distribution of late solution features are the ground-water table, geological structures such as joints, faults, folds and the nature of the lithology through which subsurface water flows.

The ground-water table is related to the prevailing relief and climate but not to the environment of deposition of the original sediment. Thus the zone of solution is not necessarily related to the depositional framework. Textural differences of the original sediments, however, may play a part in influencing the flow of phreatic waters. Tectonic fractures form important avenues for phreatic flows during late diagenesis, since lithified carbonates are far less permeable than newly deposited sediments.

Late and deep diagenetic solution might create solution porosity in limestones. Evidence of such late diagenetic solution could include leached stylolites. Water flowing through limestones becomes saturated quickly with respect to calcium carbonate; that flowing through sandstones, conglomerates, dolomites, etc., is not necessarily saturated. Thus extensive solution may occur if a limestone is interbedded between two quartz sandstone aquifers or if an unconformable surface in a carbonate is overlaid by a siliciclastic sandstone or conglomerate.

Descending phreatic waters precipitate calcium carbonate, and ascending waters might take up more dissolved calcium carbonate (see pp. 182–183) and hence create leached porosity. Pressure solution, however, eliminates pore space in limestones. The coarsely crystalline blocky cements in limestones are frequently the result of descending phreatic flows. Little is known about the exact time of formation of late diagenetic cement or about its distribution. As such cement is pore destroying it might pose a critical reservoir problem in some areas, and it is worthwhile to attempt to establish timing with the aid of other geological indications. After blocky cement formation, porosity and permeability are frequently so greatly reduced that no effective potential reservoirs remain, and water movement comes virtually to an end. The phreatic water zone has then been passed, and henceforth it is better to refer to the water as formation water.

Solution of Anhydrite and Gypsum. Molds of leached anhydrite may form an important porosity type in carbonates. Such porosity can be expected (1) near the depositional interface and (2) in areas of active hydrodynamic movement. At depths of more than a few thousand feet the formation water is undersaturated with respect to calcium sulfate. There flushing water may dissolve anhydrite. The distribution of leached-anhydrite porosity is then related to the flow path and the chemistry of ground waters.

Replacement. Dolomites can also be formed late by hydrothermal solutions or by deep formation waters. A very extensive dolomitisation by subsurface waters would imply that during a certain time interval magnesium had been moved by formation water through a rock, triggered by a hydraulic gradient. Deep formation waters are usually moving through carbonate rocks with a low average permeability if they are moving at all. Thus dolomitisation by such waters is usually limited in areal extent. Consequently the distribution of late diagenetic dolomites is usually related to some favourable paths (fractures, unconformities, etc.) that permit the flow of deep formation or hydrothermal waters through the otherwise poorly permeable or impermeable carbonates.

Summary and Nature of Diagenetic Processes

The diagenetic processes discussed here are only a fraction of those that are active but they are important for the oil geologist because they modify pore textures. Some create, others eliminate, pore space. Acquiring an understanding of such processes as dissolution, solution, leaching, precipitation and replacement may lead to improved prediction of the distribution of diagenetic products. There are two approaches:

1. Deductive, starting with a theoretical postulate which has certain implications for the distribution of diagenetic changes ("theoretical models").
2. Inductive, by review of existing and accumulation of new information on the distribution of diagenetic products; some empirical rules of thumb are arrived at ("empirical models").

In practice these approaches are closely related, and a reference model is arrived at that aids in extrapolation. Such a model presupposes an understanding of diagenesis and of a depositional mechanism. With the help of such a model the prediction of porosity distribution on the basis of existing porosity data can be attempted.

The defined diagenetic processes and their products will not be individually discussed; many good textbooks are available (e.g., Bathurst, 1971; Milliman, 1974; Flügel, 1982) in which this is done. But from studies it is obvious that certain processes are characteristic of certain diagenetic environments or zones; consequently certain diagenetic products define diagenetic facies that reflect those environments (see Fig. 3–8). Diagenetic paragenesis defines ten diagenetic environments and settings (Fig. 3–2). The definition of these terms forms the bulk of the following section.

Definitions 301–305 (Diagenetic Textural and Structural Processes)

Diagenetic processes affect carbonate particles, matrices, pores and bodies. Broadly speaking, the effects of such processes can be subdivided into textural, structural and mineralogical modifications. In each of these groups of modifications there appears to be a direct relationship between the nature and/or degree of modification and the amount of overburden. In the following grouping of terms (related to diagenetic processes only) processes giving rise to textural modifications are not separated from those leading to structural modifications, as some processes lead to textural as well as to structural products. The group of processes resulting in mineralogical modifications is separately arranged and defined, however. In this section no symbols are suggested, because there seems to be very little need for symbols indicating processes.

301. LEACHING; DISSOLUTION: the destabilization or selective removal of aragonite or high-magnesium calcite or other soluble constituents from a rock or ore body by the natural action of percolating water (see Fig. 3–9).

Note: See pp. 181–182, and Definition 438.

NE oplossing; uitloging
DE Auslaugung; Auslösung
ES disolución por percolación; lixiviación
FR lessivage; dissolution selective

302. WEATHERING: the physical disintegration and chemical and biogenic decomposition of rock that produces an in-place mantle of waste and prepares sediments for transportation. Most weathering occurs at the surface, but it may also take place at considerable depths.

NE verwering
DE Verwitterung
ES meteorización
FR altération météorique; altération superficielle

303. CHEMICAL WEATHERING: the process of weathering by which chemical reactions (e.g., hydrolysis, hydration, oxidation, carbonation, ion exchange and solution) transform rocks and/or minerals into new rocks and/or new chemical combinations that will remain indefinitely without further change under conditions prevailing at or near the earth's surface (e.g., the solution of calcium carbonate in limestones by carbonic acids derived from rainwater containing carbon dioxide).

DIAGENETIC ENVIRONMENT	PROCESS	PRODUCT	EFFECT ON RESERVOIR	TERRESTRIAL	MARINE
SURFACE-RELATED OR VADOSE DIAGENETIC (1 + 5, Fig. 3-2) (TERRESTRIAL AND MARINE)	Leaching, dissolution, solution	Mould, pore	+	T	M
		Corroded surface	±	T	M
	Encrustation	Crust	−	(T)	M
		Pisolith	−	(T)	M
	Micritisation	Micrite rim	−	(T)	M
	Bioerosion	Rhizoconcretion	−	T	(M)
	Internal sedimentation	Internal sediment	−	T	M
		Crystal/quartz silt	−	T	(M)
	Chemical solution	Fissure, breccia	+	T	M
	Vadose compaction	Unsutured grain Contacts	−	T	(M)
	Cementation	Acicular cement	−	T	M
		Point contact cement	−	T	M
		Uneven cement	−	T	(M)
		Meniscus cement	−	T	(M)
		(Micro)stalactite cement	−	T	M
		Drusy cement	−	(T)	M
		Blocky cement	−	T	M
	Mineralogical stabilisation	Aragonite	±		M
	Dolomitisation	Dolomite	+	T	M
	Dedolomitisation	Calcite pore	+ or ±	T	
	Dessication	Dessication cracks	+	T	
		Keystone vug	+	T	
		Birdseyes	+	T	M
	Bioturbation	Burrows	±	(T)	M
	Overpacking		−	T	M
	Boring	Boring, hardground	±	T	M
	Physicomechanical expansion	Tepee structure	+	T	M
	Anhydritisation	(Nodular) anhydrite	−	T	(M)
SHALLOW BURIAL (12, Fig. 3–2) Terrestrial dynamic	Leaching, dissolution, solution	Caves, pores, moulds	+		
	Cementation	Even cement, drusy	−		
		Even cement, blocky	−		
	Stabilisation	Calcite	±		
		Dolomite	+		
		Dedolomite	−		
SHALLOW BURIAL (16, Fig. 3–2) Marine dynamic	Boring	Borehole (hardground)	±		
	Burrowing	Burrow	+		
	Internal sediment	Geopetal infill	−		
	Physicomechanical expansion	Tepee structure	+		
	Dolomitisation	Dolomite	+		
DEEP BURIAL (3, 4, 7, 8, 9, 10, Fig. 3–2)	Cementation	Mainly blocky cement (zoned?)	−		
	Fracturing	Fracture	+		
	Dolomitisation	Dolomite	+		
	Pressure solution	Horsetail	−		
		Stylolite	−		
	Nodulation/differential Compaction	Nodules	−		

Fig. 3–8. Paragenesis of some diagenetic events in various diagenetic environments.

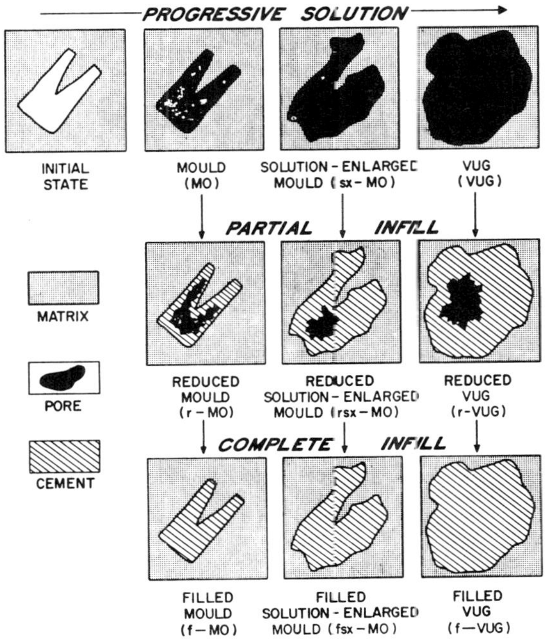

Fig. 3–9. Diagrammatic summary of steps in solution and precipitation of calcium carbonate by fresh water.

Note: See also Definitions 397, 405–408, 417 and 442.

NE chemische verwering
DE chemische Verwitterung
ES meteorización química
FR décomposition chimique; altération chimique

304. MECHANICAL WEATHERING (PHYSICAL WEATHERING): the process of weathering by which physical forces (e.g., frost action, crystal growth) break down or reduce a rock into smaller fragments, not involving chemical change.

Note: See also Definitions 412–417.

NE mechanische verwering; fysische verwering
DE physikalische Verwitterung; mechanische Verwitterung
ES meteorización física
FR désagrégation physique; altération physique

305. HALMYROLISIS: all chemical rearrangement, replacement and weathering processes operating in the subaquatic depositional environment.

Note: See also Definitions 336 and 406.

NE halmyrolise
DE Halmyrolyse
ES halmirolísis
FR halmyrolyse

306. BIOEROSION: all erosive processes being a direct consequence of faunal activity, operating on sedimentary particles, rocks or structures (see Fig. 3–10).

Note: See also Definition 406.

NE bioerosie
DE Bioerosion
ES bioerosión
FR bioérosion; biocorrosion

307. MICRITIZATION: the process of repeated boring, followed by vacating of the bore and filling of it with micrite (see Fig. 3–11). It also pertains to a process of chemical grain diminution and hence to a recrystallisation process.

NE mikritisatie
DE Mikritisierung
ES micritización
FR micritisation

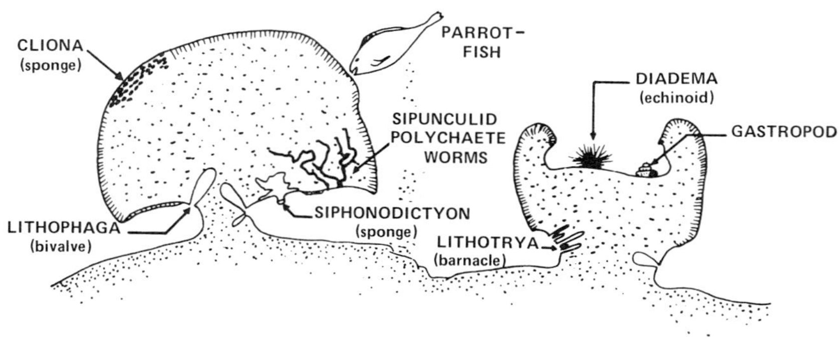

Fig. 3–10. Common organisms that erode coral reefs and their preferred location. *Lithophaga* sp. and *Lithotrya* sp. create finger-sized borings; *Diadema*, sp. and gastropods rasp, while *Cliona* sp. and *Siphonodictyon* sp. create irregular vugs and tunnels. Parrotfish crunch corals. Illustration after Ginsburg and James (1974) with their kind permission.

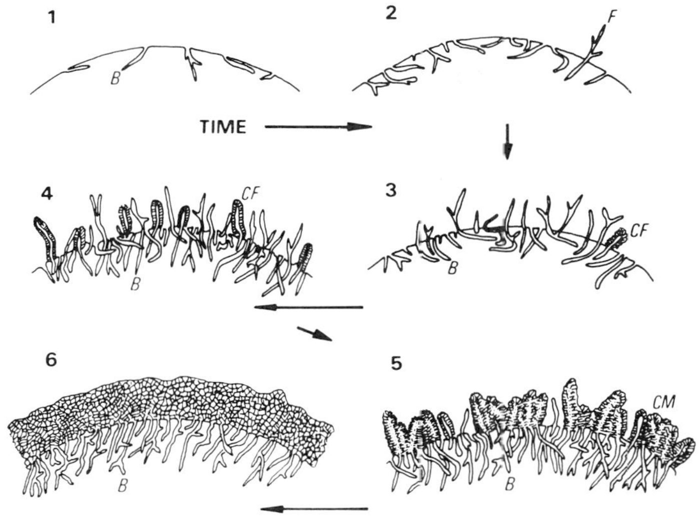

Fig. 3–11. Sequence of processes and products during generation of micritic envelope. B, Boring; F, uncalcified filament; CF, calcified filament. By Kobluk and Risk (1977). Illustration by permission of Soc. Econ. Palaeontol. Mineral.

308. CHALKIFICATION: the process whereby diagenetically stabilised shallow marine carbonates are brought into diagenetic environments (frequently exposure), in which they react with formation fluids (e.g., meteoric water), which results in destabilisation, localised leaching and extensive secondary porosity and consequently a pseudochalky rock.

Note: See also Definitions 22, 23 and 25.

NE verkrijting
DE Verkreidung
ES chalquificación
FR altération pulvérulente (des calcaires); altération crayeuse

309. CEMENTATION: a lithification process by which the pore space (primary and/or secondary) becomes partially or entirely filled by an authigenic mineral (cement).

Note: The result of cementation is cements; see Definition 340, Fig. 3–18 and pp. 203–214.

NE cementatie; verkitting; (op)vulling
DE Zementation
ES cementación
FR cimentation

310. INTERGRANULAR CEMENTATION: the formation of cement in interstices between sedimentary grains, resulting, in outward growth of crystalline material adhering to that surface, for example, the growth of calcite in the pores of an unconsolidated sand.

Note: See also Definitions 354–365.

NE tussenkorrelige cementatie
DE intergranulare Zementation
ES cementación intergranular
FR cimentation intergranulaire

311. RIM CEMENTATION (OVERGROWTH CEMENTATION; SYNTAXIAL CEMENTATION): growth of crystals in optical continuity to monocrystalline skeletal fragments.

Note: See also Definitions 369–371. For a discussion on its formation, see Walker and Barry (1984).

NE randcement; overgroeiïngscement; syntaxiaal cement
DE syntaxiale Zementation; Rindenzementation
ES sobrecrecimiento sintaxial
FR cimentation par nourrissage syntaxique; cimentation par croissance syntaxique; ciment syntaxique

312. DECEMENTATION: removal of cement by the solution process.

NE cementontrekking; decementatie
DE Lösung des Zements
ES decementación; disolución del cemento
FR dissolution du ciment

313. GRAIN GROWTH: refers to unstrained crystals that grow at the expense of their neighbours.

Note
1. See also Definitions 346, 366–368, 372 and 373.
2. Grain growth also results in newly formed crystals in places where there was other matter previously. Such crystals are neomorphic and the result of recrystallisation (Definition 318).

NE korrelvergroting
DE Kornwachstum
ES crecimiento granular
FR croissance granulaire

314. VADOSE COMPACTION: the process whereby grains are more densely packed (with fitted grain contacts) than is usually observed after normal deposition of Recent surface beach rock. Such denser packing results frequently from selective solution at grain contacts as a result of undersaturated capillary water.

Note: See also Definitions 402 and 411.

NE vadose samendrukking
DE vadose Kompaktion
ES compactación vadosa
FR compaction vadose

315. MECHANICAL COMPACTION: volume reduction of a sediment exclusively controlled by physical processes (e.g., grain rearrangement and cracking). Such volume reduction occurs mainly at the expense of pore volume.

Note: See also Definitions 276, 391–394 and 410.

NE mechanische compactie
DE mechanische Kompaktion
ES compactación mecánica
FR compaction mécanique

316. CHEMICAL COMPACTION: volume reduction of a sediment or rock that is entirely or partly controlled by dissolution processes (e.g., vadose compaction and pressure solution). Such volume reduction could occur at the expense either of pore volume or of rock volume.

Note: See also Definitions 276, 395, 396, and 402.

NE chemische compactie
DE chemische Kompaktion
ES compactación química
FR compaction chimique

317. LOCAL-SOURCE RECRYSTALLISATION; INCONGRUENT DISSOLUTION; STATIC STABILISATION: process that occurs as a result of the difference in solubility between aragonite and calcite, whereby solutions without Mg^{2+} may preferentially dissolve aragonitic particles and calcite may preferentially be precipitated. Such a process continues until all aragonite has changed to calcite.

Note: See Friedman (1964).

NE rekristallisatie van nabij
DE selektive Rekristallisation
ES recristalización de origen local
FR recristallisation sélective

318. RECRYSTALLISATION: pertaining to any change in the fabric of a mineral or a monomineralic sediment without change in chemical composition (see Fig. 3–12).

NE rekristalisatie
DE Rekristallisation
ES recristalización
FR recristallisation

<table>
<tr><th rowspan="2">PROCESS</th><th rowspan="2">EXAMPLE</th><th rowspan="2">METAMORPHIC OR IGNEOUS TERM</th><th rowspan="2">METALLURGICAL TERM</th><th rowspan="2">BATHURST (1958) VOLL (1960) ETC.</th><th colspan="4">Sholle (1978)</th></tr>
<tr><th>LOOSE</th><th colspan="3">STRICT USAGE</th></tr>
<tr><td>ONE MINERAL REPLACES ANOTHER OF A DIFFERENT COMPOSITION</td><td>CALCITE → DOLOMITE, PYRITE, ETC.</td><td>REPLACEMENT OR METASOMATISM</td><td>————</td><td>REPLACEMENT</td><td colspan="4">REPLACEMENT</td></tr>
<tr><td>A MINERAL IS REPLACED BY ITS POLYMORPH</td><td>ARAGONITE MUD OR SKELETON → CALCITE MOSAIC</td><td>INVERSION OR TRANSFORMATION</td><td>ALLOTROPIC RECRYSTALLISATION</td><td>NOT SPECIFIED</td><td rowspan="3">REPLACEMENT (LOOSELY)</td><td rowspan="2">RECRYSTALLISATION (LOOSELY)</td><td>INVERSION</td><td rowspan="3">NEOMORPHISM</td></tr>
<tr><td>A DEFORMED MINERAL CHANGES TO A MOSAIC OF UNDEFORMED CRYSTALS OF THE SAME MINERAL</td><td>STRAINED CALCITE → UNSTRAINED CALCITE</td><td>RECRYSTALLISATION</td><td>(STRAIN) RECRYSTALLISATION</td><td>RECRYSTALLISATION</td><td>STRAIN RECRYSTALLISATION</td></tr>
<tr><td>AN UNDEFORMED MINERAL CHANGES ITS FORM, GRAIN SIZE OR ORIENTATION</td><td>CALCITE MUD OR FIBERS → CALCITE MOSAIC, ETC.</td><td>RECRYSTALLISATION</td><td>GRAIN GROWTH</td><td>GRAIN GROWTH (OR GRAIN DIMINUTION)</td><td colspan="2">RECRYSTALLISATION (AND DEGRADING RECRYSTALLISATION)</td></tr>
<tr><td>A MINERAL DISSOLVES, LEAVING A CAVITY; CAVITY IS FILLED LATER</td><td>SKELETON → CAVITY → CALCITE</td><td>RARE; NO SPECIAL TERM</td><td>————</td><td>NO SPECIAL TERM</td><td colspan="4">SOLUTION – CAVITY FILL</td></tr>
</table>

Fig. 3–12. Comparison of terminology for mineral alterations as observed with petrographic microscope. Illustration by permission of AAPG.

319. INTERNAL SEDIMENTATION: a collective term referring to mechanical and/or chemical accumulation of allochthonous clastic or chemical sediments derived from the surface or from within the rock framework in cavities within the sedimentary rock.

Note: See also Definitions 387–390.

NE interne afzetting
DE Internsedimentation
ES sedimentatión interna
FR sedimentation interne

320. SHALLOW BURIAL PLASTIC DEFORMATION: volume reduction resulting in fitted contacts and distorted grains, for example, in comparatively "soft" pellets and ooids by ruptureless distortion in response to mild compressive stresses.

Note: See also Definitions 380–383.

NE ondiepe plastische vervorming
DE plastische Deformation unter geringmächtiger Sedimentbedeckung
ES deformación plástica de poca profundidad
FR déformation plastique précoce

321. DEEP BURIAL PLASTIC DEFORMATION: volume reduction resulting in lineations, fitted contacts, elongated grains and, for example, twinned calcite in compacted sediments or rock in response to elevated pressure–temperature conditions.

Note: See also Definitions 376 and 384.

NE diepe plastische vervorming
DE plastische Deformation unter mächtiger Sedimentbedeckung
ES deformación plástica profunda
FR déformation plastique d'enfouissement profond

322. PRESSURE SOLUTION: volume reduction of a rock by dissolution processes resulting from uniaxial, linear compressive stresses (e.g., overburden pressures).

Note: See also Definitions 377–379 and 383.

NE drukoplossing
DE Drucklösung

ES disolución por presión
FR dissolution par pression

323. INTRASTRATAL SOLUTION (DIFFERENTIAL SOLUTION): a solution confined to one bed or layer.

Note: See also Definition 417.

NE oplossing van één laag
DE differentielle Lösung; interstratale Lösung
ES disolución diferencial
FR dissolution différentielle

324. DIFFERENTIAL COMPACTION: local variations in degree of compaction in a sediment or rock, controlled by erratically and randomly distributed lenses of one lithology in another.

Note: See also Definitions 376 and 384.

NE differentiele samendrukking
DE differentielle Kompaktion
ES compactación diferencial
FR compaction différentielle

325. COATING: the process whereby organic tissues and/or precipitates form around a clast, grain or rock surface.

Note
1. For cortex formation in oncoids see Dahanayake (1977).
2. For a comprehensive review of many types of coated grains see the book edited by Peryt (1983).
3. See also Definitions 399, 401 and 403.

NE inkapselen; omhullen
DE Rindenbildung
ES revestimiento; recubimento
FR revêtement; enduit

Definitions 326–339 (Diagenetic Processes Leading to Mineralogical Modifications)

Calcium carbonate, $CaCO_3$, occurs as a stable (calcite) and a metastable (aragonite) mineral. The metastable modification may change to the stable one by recrystallisation. Depending on conditions in the environments of deposition or diagenesis (pH and Eh variations), other types of mineralisation may also occur. Dolomitisation holds a special place in carbonate sedimentology, as argued before. There are a series of hypotheses concerning why and when dolomitisation took place and the fabric of the resulting individual dolomite crystals or of the dolomitised rock mass. In this section, the various dolomitisation models are defined. These definitions are followed by those of the modifications in which dolomite occurs as well as some other mineralisations.

326. DOLOMITISATION: the replacement of calcium carbonate by calcium magnesium carbonate in a limestone (see Fig. 3–13).

Note: See also Definitions 9–11, 348–350, 375 and 418.

NE dolomitisatie
DE Dolomitisierung
ES dolomitización
FR dolomitisation

327. DOLOMITISATION MODEL: hypothetical system of processes by which, or settings within which, the replacement of calcium carbonate by calcium magnesium carbonate can be explained.

NE dolomitisatie model
DE Dolomitisierungs Modell

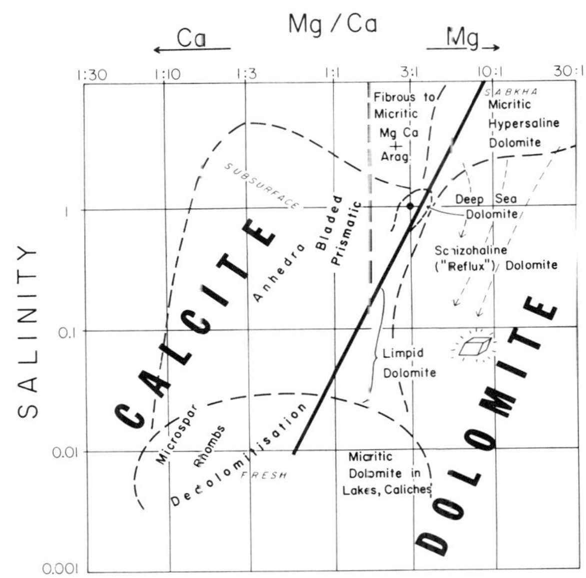

Fig. 3–13. Minerals and their morphology during dolomite formation. Folk and Land (1975). Illustration by permission of AAPG.

ES modelo de dolomitización
FR modèle de dolomitisation

328. LOCAL-SOURCE DOLOMITISATION: dolomite formation by addition of Mg^{2+} to calcium carbonates in a closed system.

NE dolomitisatie van nabij
DE lokale Dolomitisierung
ES dolomitización de origen local; dolomitización en sistema cerrado
FR dolomitisation autochtone

329. DISTAL-SOURCE DOLOMITISATION: dolomite formation by addition of CO_3^{2-} and Mg^{2+} from an outside source.

NE dolomitisatie van veraf
DE lokale Dolomitisierung durch Zufuhr von Mg^{2+} von aussen
ES dolomitización de origen distal; dolomitización en sistema abierto
FR dolomitisation d'apport; dolomitisation allochtone

330. EVAPORATIVE-PUMPING DOLOMITISATION MODEL: dolomitisation caused by upward migration of hypersaline brines due to surface evaporation.

Note: Se Hsü and Siegenthaler (1969).

NE evaporatie-pomp dolomitisatie model
DE Evaporitive-pumping Dolomitisierung Modell
ES modelo dolomitización por evaporación-bombeo
FR dolomitisation par pompage; dolomitisation par évaporation

331. EVAPORATIVE-REFLUX DOLOMITISATION MODEL: dolomitisation by downward seepage of hypersalline brines which have been formed by evaporation caused by density differences in the waters below lagoons and supratidal flats.

Note: See Adams and Rhodes (1960).

NE evaporatie-reflux dolomiet model
DE Reflux Modell
ES modelo de dolomitización por evaporación reflujo
FR dolomitisation par reflux

332. MIXING-ZONE DOLOMITISATION MODEL (DORAG MODEL): dolomitisation caused by mixed fresh and marine ground water.

Note: See Land (1973a,b).

NE mengzone dolomitisatie model (Dorag model)
DE Dorag Modell
ES modelo Dorag; modelo de dolomización por mezcla de aguas
FR dolomitisation de la zone de mélange

333. DEDOLOMITISATION: see Calcitisation (Definition 334).

NE dedolomitisatie; calcitisatie
DE Dedolomitisierung; Kalzifikation
ES calcitización; dedolomiticación
FR dédolomitisation

334. CALCITISATION: the process whereby minerals change to calcite ($CaCO_3$) (e.g., dedolomitisation).

NE kalkvorming; omzetting in kalk
DE Kalzifikation
ES calcitización
FR calcitisation

335. OVERDOLOMITISATION: continued growth of dolomite crystals after complete replacement of the original sediment ultimately plugs all intercrystalline porosity.

NE voortgaande dolomitisatie
DE —
ES supradolomitización
FR holodolomitisation

336. GLAUCONITISATION: the process of glauconite formation and/or of becoming glauconitised.

Note: See also Definition 305.

NE glauconitisatie
DE Glaukonitisierung
ES glauconitización
FR glauconitisation

337. CALCIFICATION: biogenic calcite mineralisation, for example, shell formation (see Fig. 3–14).

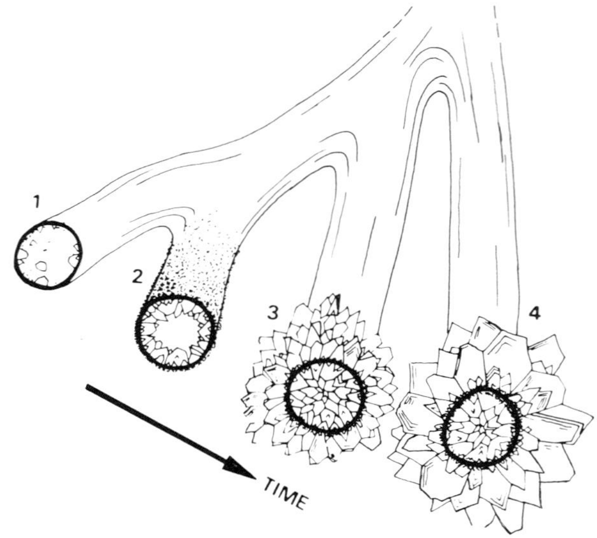

Fig. 3–14. Sequence of calcification of an exposed filament. 1, Filament dies; crystal growth begins at the interior. 2, Interior filled; exterior fully calcified by minute crystals. 3, Exterior fully covered. 4, Exterior crystallisation goes on; width of mass 30 μm.

Note
1. Skeletogenesis is *in vivo* calcification.
2. For an explanation and discussion of the process of calcification of algae filaments see Kobluk and Risk (1977).
3. See also Definition 399.

NE calcifikatie; verkalking
DE Verkalkung
ES calcificación
FR calcification

338. FOSSILISATION: the process of fossilising or becoming fossilised.

NE fossilisatie
DE Fossilisation
ES fosilización
FR fossilisation

339. GYPSIFICATION: the formation of gypsum by transformation from another substance, especially anhydrite.

NE gipsvorming; vergipsing
DE Bildung von Gips
ES yesificación
FR gypsification

3–4 DIAGENETIC PRODUCTS

Introduction

Textural Modifications

The texture of (carbonate) rocks is the spatial arrangement of compositional elements such as particles, matrix and pores. Depositional textures (e.g., see Dunham, 1962) are primary. Post-depositional rearrangements are diagenetic and secondary.

Significant texture-modifying processes are dissolution, leaching, solution, and cementation, and the grouping of definitions is consequently centered around these processes. The first three processes lead to corroded and dissolved particles and carbonate masses and to a good deal of texture-related modified primary and secondary porosity, whereas cementation may occlude primary and secondary porosity.

Leaching

A host of (bio)chemical processes may lead to dissolution of certain minerals, resulting in corroded and (partly) dissolved particles. Most commonly this occurs close to the submarine or subaerial surface. Thus leaching is usually a process active in the shallow burial diagenetic environment, but it has also been recognised in the deep burial diagenetic environment, where it enlarges, for example, stylolites and sutured contacts of interpenetrated grains. Pores that are enlargements of these stylolites and sutured contacts cannot be attributed to shallow burial diagenetic environments. It has been speculated that aggressive ground water (e.g., charged with CO_3^{2-} as a result of hydrocarbon generation) is responsible for such deep burial diagenetic leaching phenomena. Porosity is discussed in Chapter 4.

Cementation

Cementation occurs when primary and/or secondary pore space is partially or entirely filled by authigenic minerals. In carbonates this is mostly calcite, but dolomite, anhydrite and even halite are also known as cements. The resulting cement can be described in terms of crystal morphology or in relation to the substrate to which it is attached. The descriptive terminology is discussed in a subsequent section of this chapter, in which these two parameters are used simultaneously.

Diagenetic Textural Modifications

Size, shape and arrangement of crystals of diagenetic minerals in a carbonate, as well as porosity, permeability and pore type of carbonate rocks are often diagenetic textural parameters. They occur overprinted on depositional textures, and their occurrence may be influenced by the original (now perhaps relict) texture. Such a relict texture could be a depositional (see p. 28) or an earlier diagenetic one. In some instances diagenetic events completely obliterate the depositional texture, which can then be deduced only after thorough field and laboratory examination.

Textural Parameters

Size. Crystals of diagenetic minerals range in size from a few micrometers to a few millimeters. Crystalline carbonates are classified on the basis of crystal size, and the system adopted here is that of Folk (1962). The crystal size of diagenetically altered carbonates co-determines the petrophysical properties.

"Sorting". Sorting describes the grain-size distribution of transported grains. A similar term does not exist for crystal-size distribution; several terms are used to describe the size distribution of crystals (see Fig. 1–8).

A crystalline carbonate composed of crystals of about the same size is equicrystalline (even grained, equigranular or granoblastic). Different sizes that grade into each other are seriate textures, and different-sized crystals with a discontinuous crystal-size variation are referred to as hiatal textures. Porphyroblastic (hiatal) textures are characterised by the presence of large crystals in a fine matrix, whereas poikiloblastic (hiatal) textures are characterised by the presence of small crystals in a large crystal. Equicrystalline sucrose carbonates, like well-sorted clastics, have a higher porosity and permeability than inequicrystalline carbonates.

Shape. Crystal shapes can be described in terms of the degree of perfection of the crystal form and with reference to their relative dimension in space (see Fig. 3–15). Form perfection is expressed by such as terms euhedral, anhedral and subhedral (referring to decreasing degrees of development of crystal faces). A rock composed of euhedral crystals (e.g., a dolomite rock) often has considerable intercrystalline space. In rocks composed of anhedral crystals, much if not all of the intercrystalline space is taken up by irregular, interlocking crystals. Such rocks are often tight and impermeable.

Dimensional terms are equidimensional, equant or blocky when no preferential growth direction is discernible or platy, fibrous, etc., when one or two directions of the crystal are better developed than the third.

Packing. The packing of crystals is expressed by terms describing the nature of crystal boundaries. Sucrose is used when intercrystalline pore space is significant in rocks composed of randomly oriented euhedral crystals. Mosaic, interlocking and sutured crystal boundaries form a series of textures in which the crystal boundaries express an increasing degree of interpenetration. Dolomites are often characterised by a sucrose texture; recrystallised limestones tend to have a mosaic texture.

Factors Influencing Crystal Growth

The depositional texture of a sediment reflects to a certain degree processes and conditions in the environment of deposition. Likewise, the diagenetic texture of a carbonate rock reflects the conditions (temperature, pressure, degree of supersaturation and reaction rate) at the time and in

CRYSTAL MORPHOLOGY
SHAPE

		EQUANT	BLADED	FIBROUS	NEOMORPHIC
Relation to substrate	Crust	Epitaxial	Drusy/rim cements Dogtooth Bladed Isopachous Even/uneven Meniscus/dripstone	Drusy/rim cements Acicular Fibrous Isopachous Even/uneven Meniscus/dripstone	
	Overgrowth	Overgrowth Syntaxial Overdolomite		Whiskers	
	Not apparent	Saccaroidal Idiotopic Xenotopic Hypidiotopic Euhedral Subhedral Anhedral Blocky Mosaic Granular		Whiskers	Sphaerulite Sphaerulitic cement Grumose Microsparite Microcrystalline

Fig. 3–15. Descriptive terminology of crystal shapes and crystal relation to substrate. Cement crystal morphology is a function of the growth rates of its various faces and of variation of these rates with time. Growth rate is dependent on nucleation rate, temperature, saturation state of solution, crystal structure and presence of various ions in solution. These factors are not necessarily independent. See also Badiozomani *et al.* (1977) for experimental results.

the place of diagenesis. The diagenetic textural features are often described, but commonly the processes are little understood.

The following factors influence crystal size, "sorting", shape and packing:

1. Impurities. These affect crystal growth (a) by incorporation into a growing crystal; (b) by selective adsorption on certain faces, which retards or accelerates the growth of these faces; or (c) by alteration of the free surface energies relative to one another and to the medium.
2. Hydrogen ion concentration. Large, perfect crystals can grow only from solutions with suitable hydrogen ion concentration.
3. Temperature. It is not always certain whether temperature changes themselves influence crystallisation or whether they act via solvent concentration changes.

4. Saturation. A high saturation (i.e., a saturated fluid with additional undissolved crystals) may cause mother-liquor inclusion and crystal damage. Saturation differences within a medium tend to prohibit uniform crystal growth. Stirring reduces such differences and promotes the growth of uniform crystals. Saturated fluids usually produce large crystals.

Definitions 340–375 (Cement Types)

340. CEMENT: the binding, agglutinating or filling, often crystalline mineral phase that partly or entirely occludes the pores within or in between particles, regardless of its origin or composition (see Fig. 3–15).

Note: See also Definition 309.

NE cement; kit (middel); vulling; vulmiddel; van kristallijne samenstelling
DE Zement
ES cemento
FR ciment

<table>
<tr><th>Formative Mechanism</th><th>Foundation</th><th>Grain Shape</th><th>Term</th><th>Symbol</th><th colspan="2">Terms of Bathurst (1958)</th></tr>
<tr><td rowspan="10">Directly Precipitated (P)</td><td rowspan="4">Syntaxial Overgrowth (O)</td><td>Equant; on monocrystalline nucleus</td><td>x-Monocrystalline[1] Overgrowth</td><td>$P.E^*O_m$</td><td colspan="2">Rim Cement</td></tr>
<tr><td>Equant; on polycrystalline nucleus</td><td>x-Equant Overgrowth</td><td>$P.E^*O$</td><td rowspan="7">Granular Cement (if between allochems)</td><td rowspan="7">Drusy Mosaic (inside fossils, etc.)</td></tr>
<tr><td>Bladed; on polycrystalline nucleus</td><td>x-Bladed Overgrowth</td><td>$P.B^*O$</td></tr>
<tr><td>Fibrous; on polycrystalline nucleus</td><td>x-Fibrous Overgrowth</td><td>$P.F^*O$</td></tr>
<tr><td rowspan="3">Crusts (C) oriented physically on surface</td><td>Equant</td><td>x-Equant Crust</td><td>$P.E^*C_{(w)}$</td></tr>
<tr><td>Bladed</td><td>x-Bladed Crust</td><td>$P.B^*C_{(w)}$</td></tr>
<tr><td>Fibrous</td><td>x-Fibrous Crust</td><td>$P.F^*C_{(w)}$</td></tr>
<tr><td rowspan="3">Randomly oriented</td><td>Equant</td><td>x-Mosaic</td><td>$P.E^*$</td></tr>
<tr><td>Bladed</td><td colspan="2" rowspan="2">Very rare or non-existent "bladed mosaic" may perhaps exist; "fibrous mosaic" probably non-existent</td><td colspan="2" rowspan="2"></td></tr>
<tr><td>Fibrous</td></tr>
</table>

* Add number for crystal size, e.g., $P.E_4O_m$, $P.F_2C$.
x Add crystal size term, e.g., finely bladed overgrowth (or finely crystalline bladed overgrowth); medium fibrous crust (medium crystalline fibrous crust).
[1] Monocrystals may be bladed or rarely fibrous; use "bladed monocrystalline overgrowth" if desired, $P.BO_m$, etc.

Fig. 3–16. A code for directly precipitated calcite by Folk (1965). By permission of AAPG.

341. SPARITE (CONTRACTION OF SPARRY CALCITE); SPAR: a descriptive term for the crystalline, clear interstitial component of carbonates, consisting of clear calcite or aragonite. It is coarser than micrite.

Note: See also Definition 6.

NE spariet
DE Sparit
ES esparita
FR sparite

342. SPATH (obsolete); SPAR: a term loosely applied to any translucent or transparent, lustrous and cleavable mineral, for example, Iceland spar (calcite) and fluorspar (fluorite).

NE spaat
DE Spat
ES espato
FR spath

343. SPATHIC (SPATHOSE): containing sparry calcite, usually somewhat lustrous and cleavable.

NE kalkspaathoudend
DE spätig
ES espático
FR spathique

344. SUBMARINE CEMENT: cement formed in sediment pores in marine, near-surface diagenetic environments.

Note: See Macintyre (1977). See also Definition 394.

NE zeewater cement
DE submariner Zement
ES cemento submarino
FR ciment sous marin

345. SPELEOTHEM: any secondary mineral deposit, but commonly calcite, formed in a cave or along a surface by precipitation from an aqueous solution.

Note: See Kendall and Broughton (1978). See also Definition 32.

NE grotafzetting; speleoliet
DE speleothem; Höhlen-Ablagerung; Speleolit
ES espeleotema; concreción
FR spéléolite; concrétion de cavité karstique

346. MICROSPAR(ITE): neomorphic crystal agglomerate (mosaic) with crystal diameters from 0.005 to 0.02 mm.

Note: See also Definition 313.

NE mikrospar(iet)
DE Mikrosparit
ES microsparita; microesparita
FR microsparite

347. MICROCRYSTALLINE: said of crystal aggregates in which no separate crystals can be distinguished with the naked eye.

NE mikro-kristallijn
DE mikrokristallin
ES microcristalina
FR microcristallin

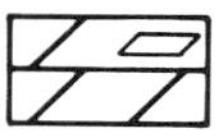

348. IDIOTOPIC DOLOMITE: said of the fabric of a crystalline dolomite in which the majority of the constituent crystals are euhedral.

Note: See also Definitions 326, 418.

NE idiotope dolomiet
DE idiotopischer Dolomit
ES dolomita idiotópica
FR idiomorphe

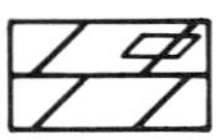

349. XENOTOPIC DOLOMITE: said of the fabric of a crystalline dolomite in which the majority of the constituent crystals are anhedral.

Note: See also Definitions 326, 418.

NE xenotope dolomiet
DE xenotopischer Dolomit
ES dolomita xenotópica
FR xénomorphe

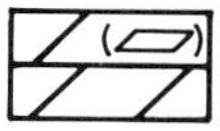

350. HYPIDIOTOPE DOLOMITE: intermediate between idiotopic and xenotopic; said of the fabric of a crystalline dolomite in which the majority of the constituent crystals are subhedral.

Note: See also Definitions 326, 418.

NE hypidiotope dolomiet
DE hypidiotopischer Dolomit
ES dolomita hipidiotópica
FR sub-idiomorphe

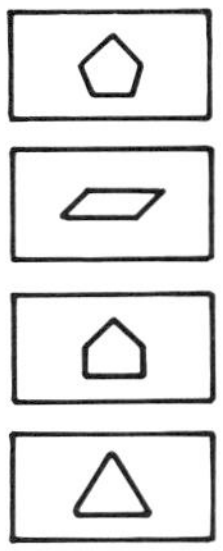

351. EUHEDRAL: said of a crystal characterized by the presence of crystal faces.

Note: See also Definition 418. See Fig. 3–15.

NE euhedraal; met kristalvlakten
DE euhedral
ES euhedral
FR idiomorphe; euhédral; automorphe

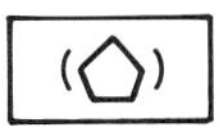

352. SUBHEDRAL: said of a crystal characterised by the presence of partially developed crystal faces.

Note: See also Definition 418.

NE subhedraal; met deels ontwikkelde kristalvlakten
DE subhedral
ES subhedral
FR subhédral

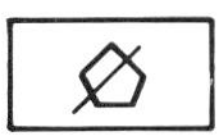

353. ANHEDRAL: said of a crystal characterised by the absence of crystal faces.

Note: See also Definition 418.

NE anhedraal; zonder kristalvlakten
DE anhedral
ES anhedral
FR anhédral

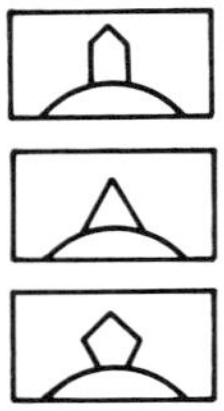

354. DRUSY (or DRUSE) CEMENT: any pore-lining cement, often consisting of radially arranged elongated crystals which do not fill the entire pore.

NE korstcement
DE Drusenzement
ES cemento de tipo drusa; cemento drúsico
FR ciment drusique; ciment géodique

355. EVEN DRUSY CEMENT (ISOPACHOUS CEMENT): cement maintaining equal thickness everywhere (see Fig. 3–17).

Note
1. For experimental even drusy cement see Badiozomani *et al.* (1977).
2. See also Definitions 357–360, 362 and 363.

NE gelijkvormig korstcement
DE gleichförmiger Drusenzement
ES cemento drúsico regular
FR ciment isopaque, cimento phréatique

356. UNEVEN DRUSY CEMENT: drusy cement not maintaining equal thickness everywhere because precipitation reflects incompletely water-filled pores (see Fig. 3–18).

Note: See also Definitions 364 and 365.

NE ongelijkvormig korstcement
DE ungleichförmiger Drusenzement
ES cemento drúsico irregular
FR ciment stalactitique; ciment vadose

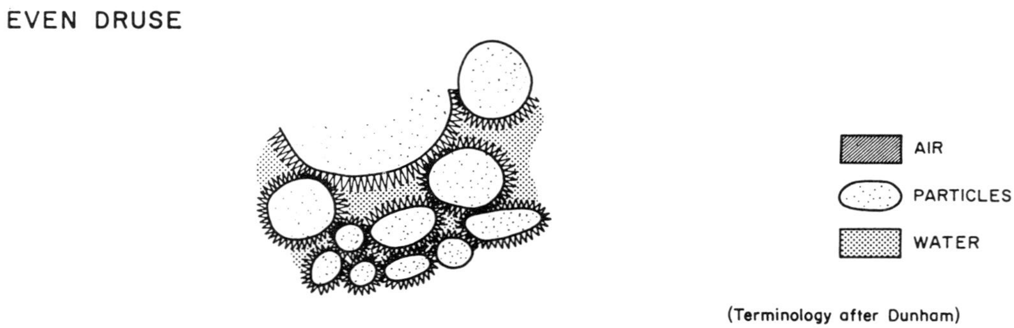

Fig. 3–17. Even drusy cements.

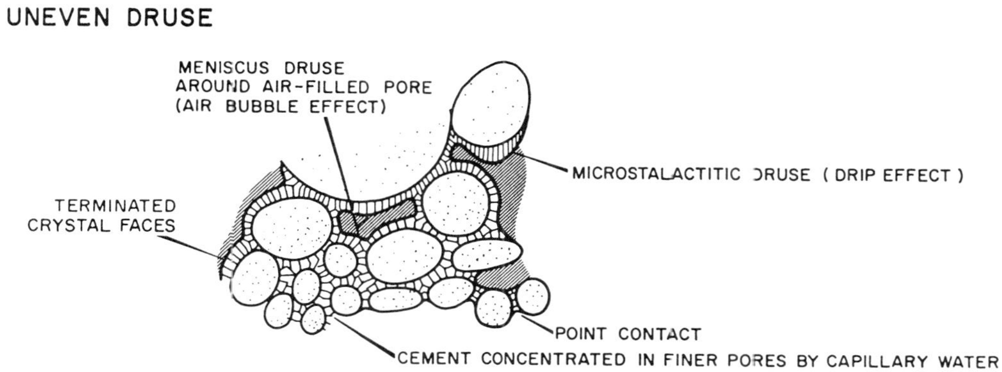

Fig. 3–18. Uneven drusy cements.

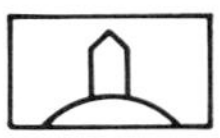

357. BLADED DRUSY CEMENT: cement consisting of crystals the length/width ratio of which grades between 6 and 1.5 (blady crystals) (see Fig. 3–19).

Note: See also Definition 355.

NE bladvormige korstcement
DE —
ES cemento drúsico en empalizada
FR ciment pallissadique

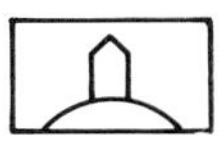

358. DRUSY DOGTOOTH CEMENT: a variety of sparry calcite cement in sharply pointed crystals of acute scalenohedral habit resembling dog teeth.

Note: See also Definition 355.

NE tandvormige korstcement
DE druzenförmiger Hundezahnzement
ES cemento drúsico en diente de perro
FR calcite "dent de chien", "dent de cochon"

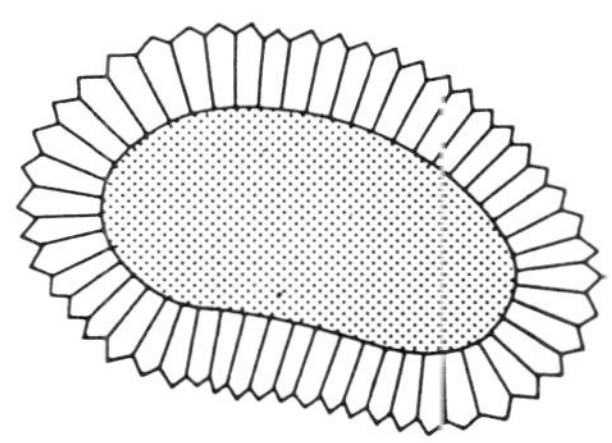

Fig. 3–19. Bladed even drusy cement. Length/width ratio, 6–1.5.

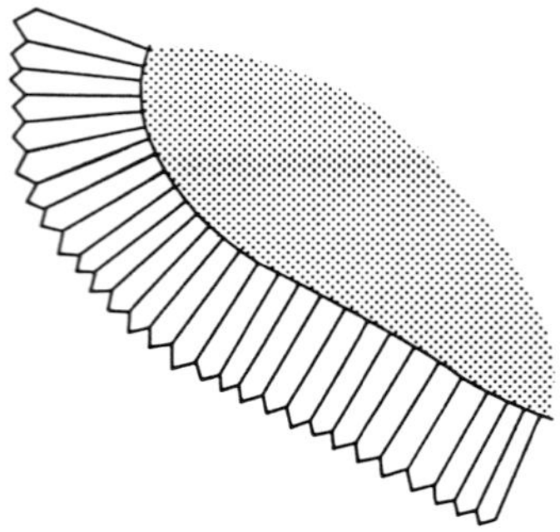

Fig. 3–20. Fibrous even drusy cement. Length/width ratio, >6.

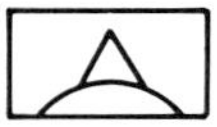

359. FIBROUS DRUSY CEMENT (CRYSTAL FIBRES): cement consisting of crystals the length/width ratio of which surpasses 6 (see Fig. 3–20).

Note: See also Definition 355.

NE draadvormige korstcement
DE fibröser Drusen-Zement
ES cemento drúsico fibroso
FR ciment fibreux

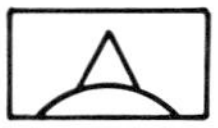

360. ACICULAR DRUSY CEMENT: cement consisting of radially arranged needlelike crystals (see Fig. 3–21).

Note: See also Definition 355.

NE naaldvormige korstcement
DE acicularer Drusenzement

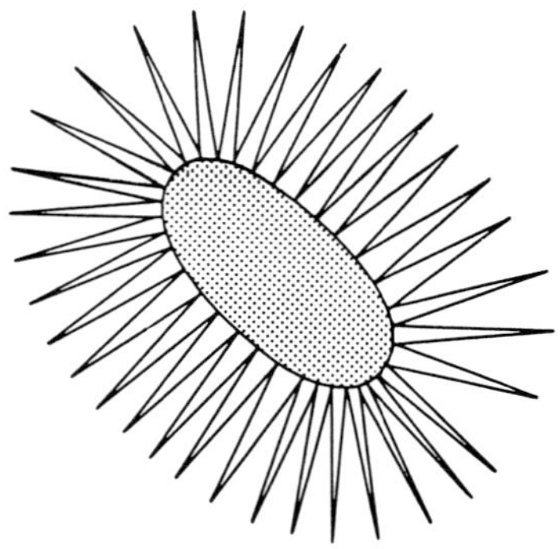

Fig. 3–21. Acicular even drusy cement. Needlelike habit, thickness diminishing along length, long or short.

ES cemento drúsico acicular
FR ciment aciculaire

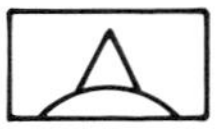

361. WHISKERS: a term for acicular cement arranged in clusters or randomly, often recognisable only on a submicroscopic scale and currently used frequently in descriptions of SEM photographs.

NE snorhaarcement
DE Whiskers-Zement
ES ''whiskers''
FR ciment aciculaire; ciment en aiguilles

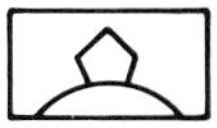

362. EQUANT DRUSY CEMENT (EQUIDIMENSIONAL DRUSY CEMENT): cement consisting of roughly equidimensional crystals (see Fig. 3–22).

Note: See also Definition 355.

NE isometrisch korstcement
DE —
ES cemento drúsico equidimensional
FR ciment granulaire; ciment à cristaux isodiamétriques

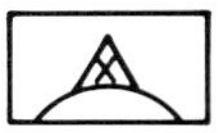

363. MICROCRYSTALLINE DRUSY CEMENT: usually the first cement surrounding grains; the crystal sizes are smaller than 0.004 mm.

Note: See also Definition 355.

NE mikrokristallijn korstcement
DE mikrokristalliner Druzenzement
ES cemento drúsico microcristalino
FR ciment drusique microscritallin

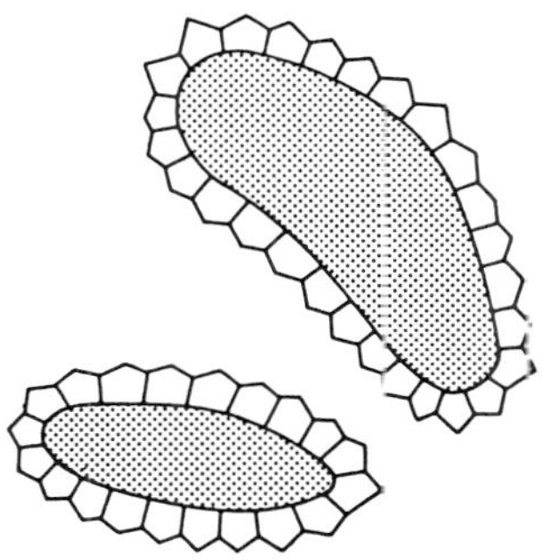

Fig. 3–22. Equant even drusy cement. Length/width ratio, <1.5.

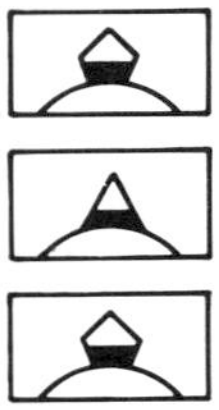

364. DRIPSTONE DRUSY CEMENT (MICROSTALACTITIC CEMENT): cement precipitated in a (partly) air-filled pore, exhibiting downward prolongations.

Note
1. For experimental results of this cement type see Badiozomani *et al.* (1977).
2. See also Definition 356.

NE druipcement (mikrostalaktitisch cement)
DE Mikrostalaktitischer Zement
ES cemento microestalactítico
FR ciment stalactitique

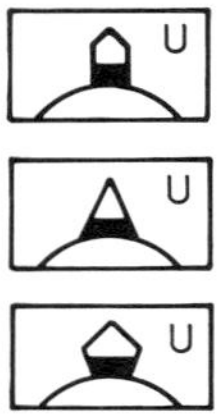

365. MENISCUS DRUSY CEMENT: drusy cement reflecting the presence of menisci in pores due to incomplete water fill.

Note
1. For experimental meniscus drusy cement see Badiozomani *et al.* (1977).
2. See also Definition 356.

NE meniscus korstcement
DE Meniskus-Zement
ES cemento drúsico con meniscos
FR ciment en ménisque

366. SPHAERULITE: a spherical body or crystalline aggregate with a radial internal structure arranged around one or more centres.

NE naaldvormig cementballetje; aggregaat; sferoliet
DE sphaeroid
ES esferolito; esferulito
FR sphérolite

367. SPHAERULITIC CEMENT: radiating, closely packed acicular cement.

NE naaldvormige cementballetjes
DE sphärolitischer Zement
ES cemento esferolitico; cemento esferolítico
FR ciment sphérolitique

368. GRUMOSE (GRUMOUS; CLOTTY): globular, irregularly shaped, vaguely defined clots made up of extremely fine crystalline calcite, standing out as dark grey in a matrix of colourless granular calcite.

Note: See also Definition 70.

NE klonterig, klonterige structuur
DE krümelig
ES grumoso
FR grumeau

369. SYNTAXIAL RIM (SYNTAXIAL OVERGROWTH): an optically oriented crystal overgrowth of a detrital grain, developed during diagenesis.

NE syntaxiale rand
DE syntaxiale Rinde
ES cemento sintaxial; sobrecrecimiento sintaxial
FR ciment syntaxial; croissance syntaxiale

370. SYNTAXIAL OVERGROWTH: formation of cement crystals in optical continuity with monocrystalline particles.

NE syntaxiale groei
DE syntaxialer Aufwuchs
ES sobrecrecimiento sintaxial
FR croissance syntaxique; nourrissage syntaxique

371. EPITAXIAL (OVER)GROWTHS: diagenetically produced overgrowths on clastic grains having a different composition (e.g., calcite on quartz).

NE epitaxiale groei
DE epitaxialer Aufwuchs
ES sobrecrecimiento epitaxial
FR croissance épitaxique

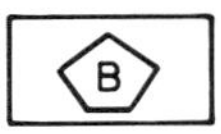

372. BLOCKY CEMENT: cement consisting of crystals characterised by their large, approximal, equidimensional size.

Note: See also Definition 341.

NE blokcement
DE Blockzement
ES cemento en bloque
FR ciment isogranulaire

373. MOSAIC CEMENT: cement composed of approximately equigranular crystals, usually with simple, straight grain boundaries (see Fig. 3–23).

NE mozaïekcement
DE Mosaikzement
ES cemento en mosaico
FR cement à cristaux en mosaïque

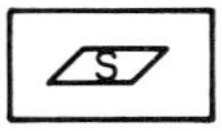

374. SACCHAROIDAL; SUGARY; SUCROSE: having a finely granular or crystalline, "sugary" texture (e.g., limestones, dolomites).

NE suikerachtig
DE Zucker-Gefüge
ES sacaroidea
FR saccharoïde

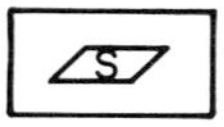

375. SUCROSE DOLOMITE (SUCROSIC DOLOMITE; SACCHAROIDAL DOLOMITE): a term commonly applied to idiotopic dolomite rock.

Note: See also Definition 326.

NE suikerachtige dolomiet
DE Zuckerdolomit
ES dolomía sacaroidea
FR saccharoïde

Diagenetic Structural Modifications

Review of Diagenetic Structures

Depositional structures (pp. 57–60) in (carbonate) rocks denote a product (the result of "construction") or a manner of organization, arrangement or interrelation of constituting parts. Diagenetic (secondary) structures denote a product that is the result of a rearrangement of the depositional organisation, the depositional arrangement or the depositional interrelation of parts of a (carbonate) rock mass.

Sometimes it is possible to relate diagenetic structures to overburden; frequently such structures are the result of direct influence of the depositional environment. A separation is here made between diagenetic structures that affect only grains or crystals and those that affect larger carbonate masses (Table 3–1). As a result of a complex of (bio)chemical, physical and mechanical processes, amongst which weathering, dissolution, solution and leaching play a significant role, major profiles or surfaces (structures) are formed after prolonged exposure in terrestrial settings in which carbonates are present. All such features are the result of weathering, solution and erosion. The climate determines mainly whether calcrete, caliche, a scalloped surface or a karst surface will be formed (going from more arid to more humid circumstances) (see Figs. 3–24 and 3–25).

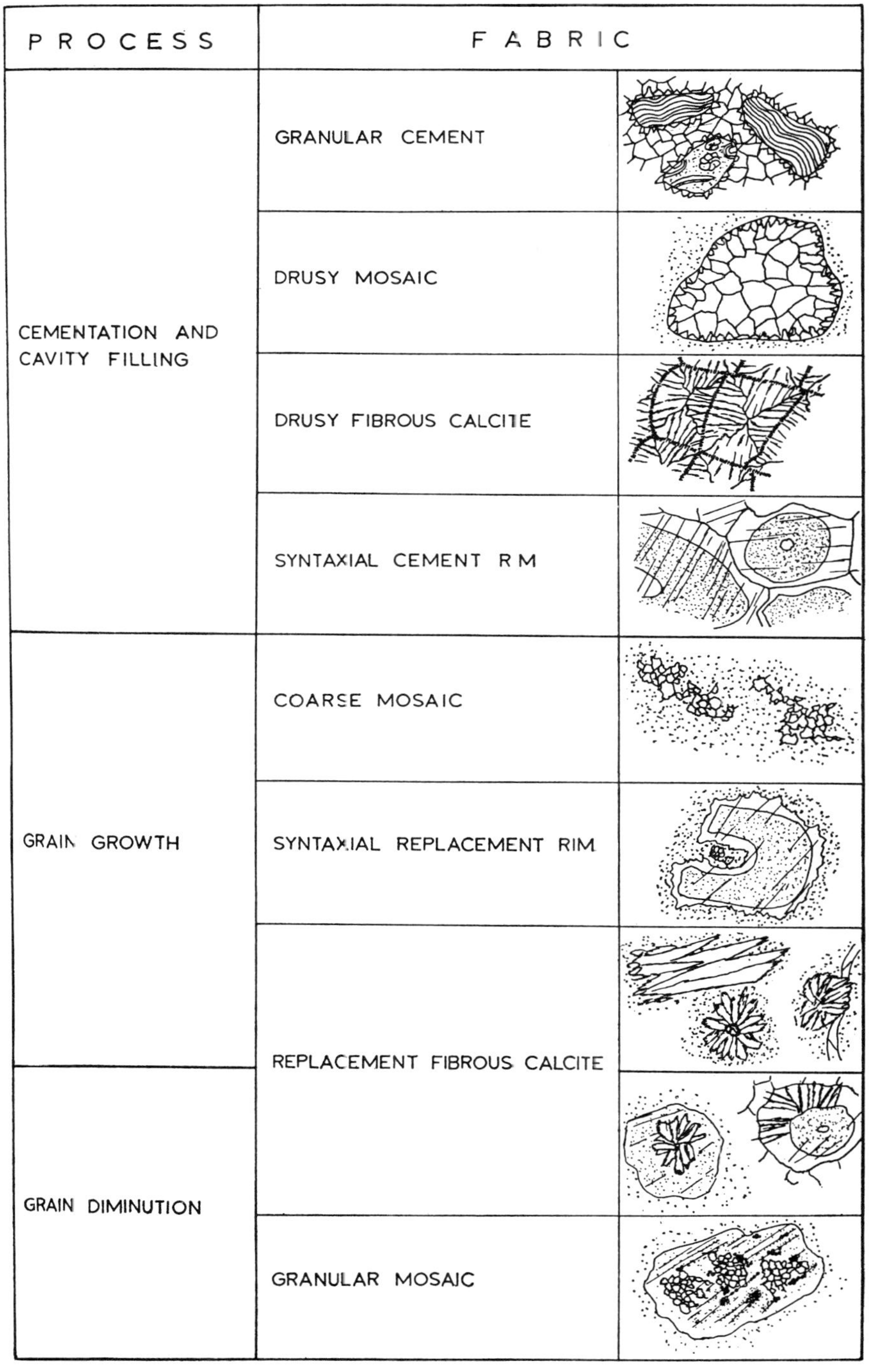

Fig. 3–23. Diagenetic fabrics in limestone. From Orme and Brown (1963). Illustration with permission of the Yorkshire Geol. Soc.

Table 3–1

Types of Diagenetic Structures in Various Diagenetic Elements

	In individual grains or crystals	In rock units
Surface related	Cortex Spalling (exfoliation)	Fracture Solution breccia Geopetal structure Birdseye structure Keystone vug structure Stromatactis Vadose silt Scalloped surface Karst surface Hardground Caliche Rizoconcretion Pseudoanticline/tepee Submarine pseudoanticline Dessication crack Flat-pebble breccia Reef tufa
Overburden related	Grain interpenetration Grain repacking Grain or crystal breakage Sutured grain contacts Faced grain contacts Fitted grain contacts Spalling (exofoliation) Distorted ooid	Horsetail Stylolite Nodule Fracture Cone-in-cone structure Enterolithic structure Interstratal contortion Intraformational Contortion Intrastratal flow structure

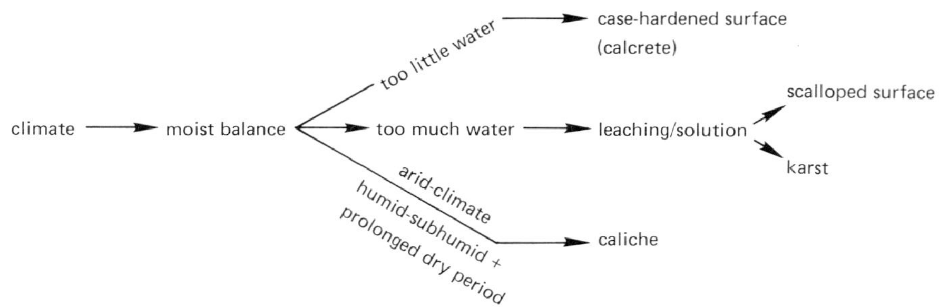

Fig. 3–24. Possible relationships among calcrete, caliche, scalloped surface and karst.

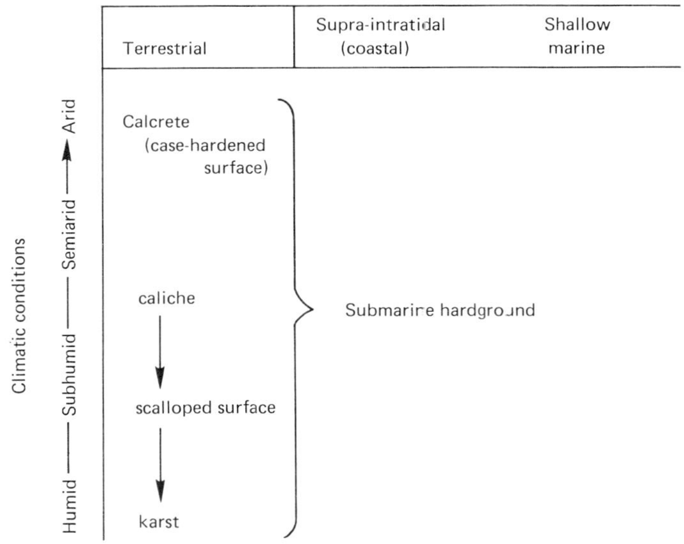

Fig. 3–25. Main settings of calcrete, caliche, scalloped surface and karst.

Types of Diagenetic Structures

The significance of depositional structures for environmental reconstruction has been recognised and exploited (see pp. 58 and 71); the use of diagenetic structures for the reconstruction of diagenetic environments, however, has still some way to go. Gradually it has come to be understood that diagenetic structures are clues that help to interpret the depositional and diagenetic histories of a carbonate sequence. They help to outline belts of potential reservoirs and may portray possible trapping conditions.

Like sedimentary depositional structures, diagenetic ones are recognised in a wide range of sizes. Some large structures are perhaps better termed *diagenetic bodies* (e.g., lithoherm). The distinction that was practical in depositional structures and grains, namely, attachment to the substrate (structures attached; grains unattached), is not practical in diagenetic structures. Consequently both particles (e.g., pisoids) and masses of particles (e.g., vadose silt) are here considered to be diagenetic structures.

In terms of their formative agents, diagenetic structures can be classified as chemical and physical. Structures with a chemical origin are (1) precipitational structures and (2) solution structures. Structures with a physical origin are (1) resedimentation structures and (2) deformational structures. The most commonly observed diagenetic structures are listed in Table 3–2.

Precipitational Structures. Diagenetic structures resulting from the formation of diagenetic minerals are referred to as precipitational. These include mineral segregations forming accretional bodies. The segregations take the form of layers, nodules, rosettes, void fillings or veins. They may replace the host rock, fill openings or form by thrusting aside the enclosed matrix. Another class of precipitational structures is formed by grains (e.g., pisoliths).

Table 3–2

Most Commonly Observed Diagenetic Structures as Products of Chemical or Physical Processes

Process	Formative mechanism	Structure
Chemical	Precipitational	Layered segregations Nodules Rosettes Void fillings; fill structures Veins Mottling Pisoids Calcite concretion; calcicrust Obliterated bedding
	Solution	Stylolites Vugs Moulds (fossil moulds, crystal moulds, oomoulds)
Physical	Resedimentation	Diagenetic sediment
	Deformational	Tepee structures Crumbly fractures Fractured, jointed and contorted beds Collapse breccias

1. Nodules. Nodules are rounded or irregular bodies of mineral matter (chert, anhydrite, phosphate) unlike that of the host rock in which they occur. Anhydrite nodules in carbonates could be diagenetically altered aggregates of selenite crystals. Such nodules have a texture similar to that of bedded anhydrite. Anhydrite and gypsum, being relatively soluble, are rarely present in outcropping rocks, but nodules of sparry calcite are found. These could be replacement products after anhydrite or gypsum nodules.

2. Rosettes. Clusters of large euhedral crystals in sedimentary rocks form rosettes that might occasionally have been formed by vadose waters, elsewhere by supersaline seepage reflux waters.

3. Void fillings. Vugs, fossil cavities or other pore space in a rock might be filled by diagenetic minerals. These occur as drusy crystals followed by blocky cement. The void fillings assume various forms and characteristics. One characteristic fill structure is stromatactis, occurring commonly in Palaeozoic carbonate mud mounds. Its origin is widely disputed. Various void-fill structures are of significance in the exploration of minerals such as "strata-bound lead zinc". Classifications could be based on the presence or absence of fill structures and on the nature of the (original) interconnection of the pores (effective porosity) (see Definition 423).

4. Veins. Material in veins in carbonate rocks is calcite, dolomite, anhydrite, gypsum, etc. Some vein material fills fractures that were produced during sedimentation (e.g., mud cracks) or during shallow burial diagenesis (e.g., crumbly fractures); others fill tectonically induced fractures (deep burial diagenetic zone). The removal of unstable minerals in veins could produce the interconnecting pore space which renders a vuggy carbonate permeable.

5. Mottling structures. The origin of mottled carbonates is related to (selective dolomitisation of) infilled cracks or of burrowed zones.

6. Pisoids. Pisoids in carbonate rocks are considered sedimentary particles, often of an early vadose diagenetic origin. Dunham (1969b) discussed the criteria for this interpretation in full. Evidence for *in situ* growth indicates that pisoids are not equal to algal oncoids; it also indicates that pisoids are not normal sedimentary particles, but some kind of concretionary structure. Practically identical pisolitic structures have been observed in Recent caliche deposits.

7. Calcretes; calcicrusts. Calcretes result from inorganic precipitation during surface-related diagenetic processes at an early or a later time in the sedimentary history. They require for their formation extensive periods of exposure. Specific climatic settings are required (see Fig. 3–24, 25).

8. Massive bedding. Some massive carbonates were originally bedded. Diagenetic processes (dolomitisation, vadose alteration) obliterated bedding features. Occasionally such massive carbonates are mistakenly called reef.

Solution Structures. Structures in rocks that are the result of solution or leaching of preexisting minerals are called solution structures. Common features are stylolites, vugs and moulds. Solution requires the removal and transport of material. An understanding of the nature of solution and the movement paths of dissolved material is a prerequisite for predicting the distribution of solution along certain structures.

1. Stylolites. The origin of stylolites is controversial. Most authors consider it a pressure solution structure. The geometric relation between stylolites and particles or structures indicates removal of considerable amounts of rock material, which requires hydrodynamic movements. Such movements and their driving mechanism have not yet been thoroughly investigated. Although solution along grain contacts in carbonates is often microstylolitic, it is not always so. Dunham noted solution features resulting in closely packed grains in unburied Quaternary sediments and showed that these features were produced by vadose waters under near-surface conditions and not by pressure solution at great depth.

2. Vugs; moulds. Vugs and moulds are pores in rocks, commonly formed as a result of solution by vadose or phreatic waters, occasionally by entrapment of air in the swash zone of a beach (keystone vugs). Some vugs were formed by fracturing and internal erosion during early diagenesis in a surface-related diagenetic zone.

Vugs may be irregular in outline; they are often lined with drusy crystals and may be partially filled with internal sediments. Moulds could approximate the shape of the dissolved particle or crystal. Vugs and moulds provide significant storage space for petroleum. Carbonate rocks with interconnected moulds are often important reservoirs.

Resedimentational Structures. Void space in carbonate rocks is often floored or filled by a fine internal sediment of marine or diagenetic origin. Marine internal sediments rest directly on the floor of voids; internal sediment of diagenetic origin (crystal silt; Dunham, 1969a) rests on drusy cement. It may be overlaid by blocky cement, which is ordinarily younger than the drusy cement. Dunham showed that such crystal silts were deposited by downward-percolating vadose waters shortly after the deposition of enclosing carbonates.

Deformational Structures. Diagenetic changes commonly involve the transport of material from one place to another. Deformation could be the result, because stress can be induced in a carbonate bed by the removal or addition of material. Amongst the features induced by compressive stresses are tepee structures or pseudoanticlines. Crumbly fractures, fractured and jointed beds, contorted beds and collapse breccias are amongst the deformational structures formed as a consequence of the removal of material.

1. Tepee structures. Small chevron folds form shortly after deposition, possibly in relation to compressive stress induced by expansion. In places the crests of the folds are eroded before deposition of the overlying beds.

2. Crumbly fractures. Lightly cemented carbonate rocks in the surface-related diagenetic zone form crumbly fractures that break around instead of across grains. Thus they differ from sharp-edged features of tectonic origin. The wide separation of the walls of some crumbly fractures suggests considerable physical erosion and leaching inside the rock. Crumbly fractures are often floored by internal sediments.

3. Jointed, faulted and contorted beds. Outcropping evaporites will usually dissolve. This could cause deformational structures, local fractures, small faults and contorted bedding in the overlying carbonates.

4. Collapse breccia. Collapse breccia consist of angular fragments of originally bedded carbonates overlying an evaporite bed that has been dissolved away. As a result of karstic solution also, collapse breccia could be formed.

Significance of Diagenetic Structures

Diagenetic structures discussed in relation to the processes that form them aid in reconstructing depositional and diagenetic history. Careful observation and logical deductions are needed to make the link between mode of genesis and resulting structure.

Both depositional and diagenetic structures occur in associations. Structure associations are characteristic of certain settings and environments, both depositional and diagenetic. Moreover, such associations are helpful for establishing the nature of diagenetic processes. Diagenetic structures understood in this way have until now been investigated only occasionally with the aim of deriving conclusions about the diagenetic environment. Thus their full significance in this respect cannot always be fully evaluated.

Definitions 376–418 (Diagenetic Structures)

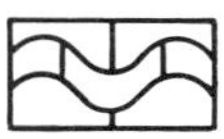

376. NODULAR LIMESTONE (LENTICULAR LIMESTONE): limestone characterised by the abundant presence of small-scale lenticular structures and associated horsetails (flasers).

Note: See Carrison and Kennedy (1977). See also Definitions 321 and 324.

NE knobbelige kalk; lensachtige kalk
DE Knollenkalk; Flaserkalk
ES caliza nodulosa
FR calcaire noduleux

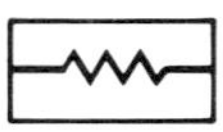

377. STYLOLITE: a complexly shaped seam or interface with a number of roughly columnar or fingerlike extensions that fit in between the columnar extension of the opposite mass (see Fig. 3–26).

Note: See also Definition 322.

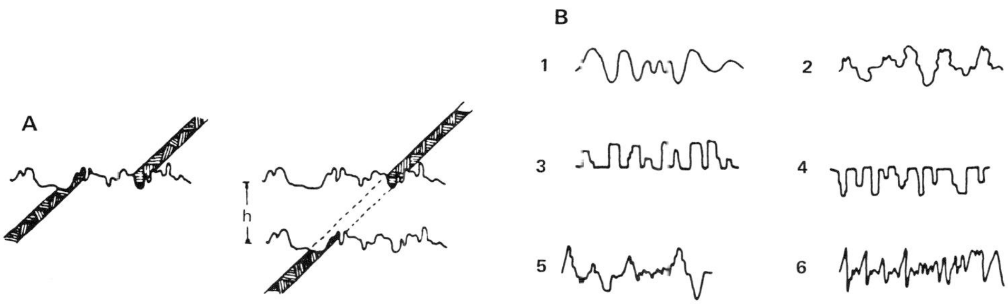

Fig. 3–26. A, Apparent offset of vein due to dissolution of interval *h* along stylolitic seam. B, Cross-sectional view of a stylolitic seam in carbonates. Types are (1) simple, (2) sutured, (3) rectangular up, (4) rectangular down, (5) sharp peak, (6) "seismogram type. Terminology after Park and Schott (1968). Illustration by permission of Springer-Verlag.

NE stylolieт
DE Stylolith
ES estilolito
FR stylolite

378. STYLOLITIC CONTACT; STYLOLITIC SEAM: interpenetrating interface within a rock.

Note: See also Definitions 322 and 379.

NE stylolitisch contact
DE stylolithischer Kontakt
ES contactos estilolíticos
FR contact stylolithique

379. STYLOLITIC: pertaining to a stylolite.

Note: See also Definitions 322 and 378.

NE stylolitisch
DE Stylolithisch
ES estilolítico
FR stylolithique

380. GRAIN INTERPENETRATION: collective term for faced, fitted and sutured contacts.

NE korrelineengroeiing
DE Korndurchdringung

ES interpenetración de los granos
FR interpénétration granulaire

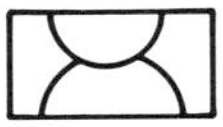

381. FITTED GRAIN CONTACT: curved interface between two grains due to one grain that dissolved, allowing the other grain to penetrate.

NE in elkaar passende korrelcontacten
DE übereinstimmende Flächenkontakte
ES contacto linéar (entre granos)
FR contact linéaire grain à grain

382. FACED GRAIN CONTACT: straight interface between two grains, due to equal solution of both grains.

NE vlak korrelcontact
DE Flächenkontakte
ES contacto rectilíneo (entre granos)
FR contact rectiligne euhédral

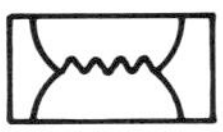

383. SUTURED GRAIN CONTACT: interpenetrating interface between two grains due to irregular solution of both grains.

NE sutuurcontact
DE suturierter Kontakt
ES contacto par sutura
FR contact par interpénétration; contact par engrenement

384. HORSETAIL STRUCTURE (FLASER; SEDIMENTARY BOUDINAGE): wispy or ramifying seams in which insoluble residues are accumulated as a result of rock-volume reduction due to differences in solubility.

Note: See also Definitions 322 and 324.

NE oplossings paardestaart
DE Flaserstruktur
ES estructura en cola de caballo
FR structure en queue de cheval

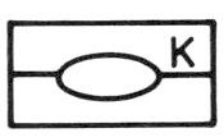

385. KEYSTONE VUG: a primary sedimentary vug, the size of which is usually larger than the size of the components (grains) of the rock.

Note: See also Definition 432.

NE sluitkorrelopening
DE Fenster-Gefüge
ES fenestrae
FR fenestrae; micro-caverne

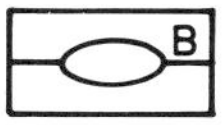

386. BIRDSEYE STRUCTURE: patches of sparry calcite in micritic carbonates, indicating supratidal environmental conditions.

Note: Apart from calcite, other minerals (e.g., anhydrite) could fill the opening in the micritic carbonate.

NE spaatogen
DE Spatitaugen
ES birdeseye (estructura en —)
FR structure oeillée; "birdseye"

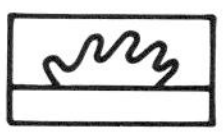

387. STROMATACTIS: open-space sedimentary structures characterised by horizontal or nearly flat bottom and irregular upper surfaces, formed by the filling with internal sediments and/or sparry calcite cement of original cavities, usually within micritic carbonate (see Fig. 3–27).

Note: See also Definition 319.

NE stromataktis
DE Stromatactis
ES estromatactis
FR stromatactis

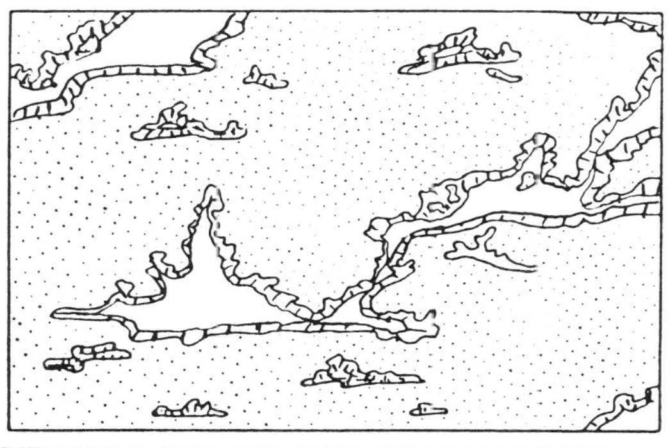

Fig. 3–27. Characteristic form of stromatactis structure with sparry calcite infill. Forms vary in dimension between 1 and 20 cm.

388. INTERNAL SEDIMENT: sediment geopetally deposited in vugs, body cavities or fractures in (semi)consolidated rocks.

Note: See Assereto and Kendall (1977). See also Definition 319.

NE intern sediment
DE Internsediment
ES sedimento interno
FR sédiment interne

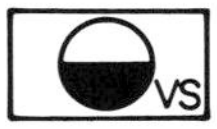

389. VADOSE SILT: a silt-sized product of internal erosion related to subaerial exposure (Dunham, 1969a).

Note: See also Definition 319.

NE vadose slib
DE Vadoser Silt
ES limo vadoso
FR silt vadose

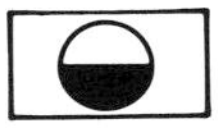

390. GEOPETAL STRUCTURE: cavity or void structure (partly) filled with sediment that indicates bottom and top during deposition.

Note: See also Definition 319.

NE aardgerichte struktuur
DE Geopetalgefüge
ES estructura geopetal
FR structure géotrope; structure géopétale; figure de polarité

391. REPACKING: mechanical grain rearrangement in response to compressive stresses, resulting in partial loss of depositional porosity.

Note: See also Definition 315.

NE korrelverspreidingsverandering
DE —
ES re-empaquetamiento
FR réarrangement mécanique

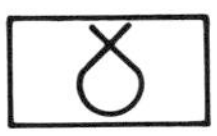

392. TEPEE STRUCTURE: expansional, non-tectonic anticline of centimetre to metre scale.

Note
1. For very detailed petrography see Assereto and Kendall (1977). See also Definition 315.

NE tepee struktuur
DE Tepee-Struktur
ES estructura en tepee
FR structure en "tepee"

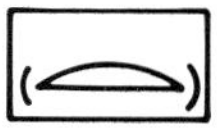

393. PSEUDOANTICLINE: a foldlike deformation structure, polygonal in plan, with an anticlinal appearance involving sheets of sediments and often separated from one another by wedges.

Note
1. The difference between tepee and calcrete pseudoanticlines has been discussed by Watts (1977).
2. See also Definition 315.

NE pseudoanticline
DE Pseudoantikline
ES seudoanticlinal
FR pseudo-anticlinal

394. SUBMARINE PSEUDOANTICLINE: buckled margins of expanded submarine megapolygons formed in carbonate beds during early submarine cementation.

Note: See Shinn (1969). See also Definitions 315 and 344.

NE onderzeese schijnanticline
DE submarine Pseudoantikline
ES seudoanticlinal submarino
FR pseudo-anticlinal sous-marine

395. CONE-IN-CONE STRUCTURE: an accumulation of cones, usually consisting of calcite, fitting into one another, the apices pointing downward (when occurring in layers) or inward (when formed in the exterior of concretions).

Note: See also Definition 316.

NE 'cone in cone' struktuur
DE Tütenmergel; Nagelkalk
ES estructura "cone-in-cone"; estructura en cono encajado
FR "cone-in-cone"

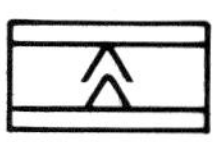

396. BEEF: a term used originally in Purbeck (England) for thin layers or lenses of secondary, fibrous calcite (or other fibrous material; e.g., gypsum), lying between layers of shale. At some places they contain cone-in-cone structures.

Note
1. See also Definition 316.
2. There are no equivalents of this term in other languages.

397. RUIN MARBLE STRUCTURE: a structure typically found in micritic limestone with very thin calcite veins, separating blocks with different patterns of iorn (hydr)oxide infiltration, that in cross section sometimes give the impression of a picture of ruins. The structure is also found in other rocks (e.g., in fine-grained sandstones).

Note: See also Definition 303.

NE ruïnen-marmer-struktuur
DE Ruinenmarmer
ES estructura ruiniforme
FR structure veinée comme un marbre

398. VERMICULAR STRUCTURE: structure containing wormlike forms.

Note: See also Definition 399.

NE wormvormige strukturen
DE wurmförmige Textur
ES estructura vermicular
FR structure vermiculaire; structure vermiculée

399. RHIZOCRETIONS: micrite-cemented root tubes that may be hollow and may consist of a succession of vague, concentric laminations of micrite; also micritic material that encloses a root.

Note
1. The vegetal tissues are never calcified.
2. See also Definitions 325, 337 and 338.

NE wortelgangen
DE Rhizokonkretionen
ES rizocreción
FR rhizoconcrétion; racine calcifié

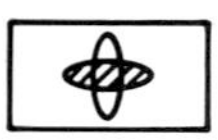

400. DOLOMITE MOTTLING: a textural feature resulting from incipient or arrested dolomitisation of limestones, characterised by preferential alternation that leaves patches, blotches, birdseyes, laminae, allochems and/or other structures unaffected; also a similar phenomenon resulting from arrested or incomplete dedolomitisation.

Note: See Kendall (1977).

NE dolomiet-vlekkig
DE dolomitische Flecken-Textur
ES manchas de dolomitización
FR mouchetage dolomitique

401. CORTEX: the covering layer or veneer of secondary material around a grain.

Note: See also Definition 325.

NE huidje
DE Schale; Rinde
ES córtex
FR cortex

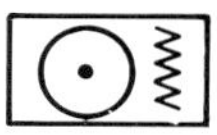

402. SPALLED GRAIN: an exfoliated grain due to weathering; often used for grains with concentric texture (e.g., ooids, particles rimmed with drusy cement).

Note: See also Definition 316.

NE afgeschilferde korrel
DE angebrochene Korn
ES grano exfoliado
FR desquamation

403. COATED: covered by a film or veneer of secondary material.

Note: See also Definition 325.

NE voorzien van een huidje; —film; ingekapseld
DE umhüllt
ES revestido; recubierto
FR revêtu (revêtir); enduit (enduire)

404. BEACH ROCK: intertidal/supratidal, loosely consolidated or well-consolidated sedimentary rock, consisting of calcareous debris and cement derived from evaporating seawater.

Note: See also Definition 314.

NE strandversteening
DE beachrock

ES "beachrock"; roca de playa
FR grès de plage

 C

405. CALICHE (CALCRETE): indurated carbonate soils occurring predominantly in semiarid to arid climates.

Note
1. See Assereto and Kendall (1977). Discussion of antiform expansion structures in shape related to tepees, but in genesis related to caliche.
2. See Knox (1977) and Harrison (1977). Barbados has a subhumid to humid tropical climate, but locally there can be prolonged dry periods (up to 5 months). Harrison lists significant petrographic fabrics which aid in recognising caliches.
3. See also Definitions 273, 276, and 303.

NE caliche
DE Caliche; Krustenkalk
ES caliche
FR croûte calcaire

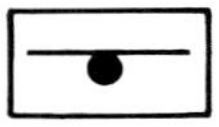

406. HARDGROUND: a bed of limestones, the upper surface of which (1) has been bored, corroded or eroded (by abrasion); (2) has been encrusted or overgrown by sessil organisms; or (3) has acted as a source for pebbles that are included in the overlying sediment layer. It indicates a time of no or low sedimentation and early cementation.

Note
1. Submarine hardgrounds are characterised by three-dimensional spongeworks created by bioeroding organisms (see Ginsburg, 1953).
2. See also Definitions 303, 305 and 306.

NE verhard afzettingsvlak; erosievlak
DE Hartgrund
ES "hardground"
FR surface durcie

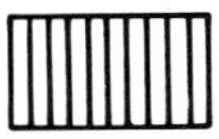 K

407. KARST SURFACE: a type of topography formed over carbonate or gypsum by solution; it is characterised by closed depressions or sinkholes, caves and underground drainage (see Fig. 3–28).

Note: See also Definitions 301–303 and pp. 181–182.

NE karstoppervlakte
DE Karstoberfläche

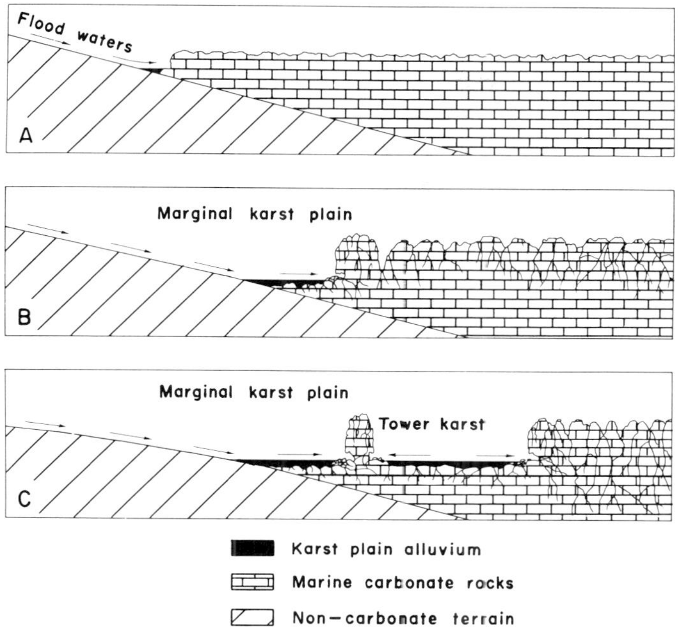

Fig. 3–28. Sequence in the development of a karst landscape. A, Flood waters begin dissolution; dissolution and solution collapse create scarps. B, C, As scarps retreat, isolated karsblocks (tower karst) may remain. If shallow sea floods such a landscape, reefs could form on top of karst remnants (see also Fig. 2–72). Illustrations by permission of J. Wiley & Sons Inc.

ES superficie carstica
FR surface karstique

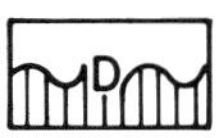

408. SCALLOPED SURFACE: smooth curved basins (10–100 cm wide) with intervening steep, locally overhanging walls and sharp ridges. Relief is up to 20 cm. Scalloped surfaces may pass laterally into planar erosional contacts. Vadose diagenetic features may be associated with scalloped surfaces.

Note: See Read and Grover (1977). See also Definition 301 and p. 182.

NE uitgehold oppervlak
DE —
ES superficié festoneada
FR micro-cuvette

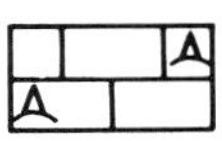

409. REEF TUFA: drusy, prismatic, fibrous calcite deposited directly from supersaturated water upon the void-filling internal sediment of the calcite mudstone of a reef knoll.

NE rif tuf; rif cement
DE Riff-Zement
ES toba arrecifal
FR ciment calcitique de vides dans un recif

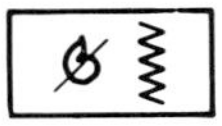

410. (GRAIN) BREAKAGE: brittle fracturing of grains in response to compressive stresses.

Note: See also Definition 315.

NE korrelbreukvorming
DE Kornbruch
ES fractura (de grano)
FR fracturation granulaire

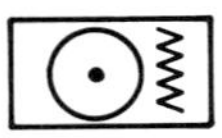

411. DISTORTED OOID: (1) a normal ooid with a distorted cortex or displaced nucleus or both; (2) an ooid with a modified external shape. Laminae of the cortex may be distorted or undistorted but interrupted by intergranular pressure solution.

Note: Ooids with distorted external shapes have been attributed to tectonic deformation, depositional processes and diagenetic post-depositional processes (see Definitions 304, 314 and 320); those with internal distortion have been attributed to diagenetic changes. For a discussion see Corley (1977).

NE vervormde ooide
DE verformte Ooide
ES ooide deformado; oolito deformado
FR oolithe déformée

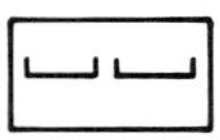

412. DESICCATION CRACK: a crack fracture as a result of repeated insolation and wetting.

Note: See also Definition 304.

NE uitdrogingsbreuk
DE Schrumpfriss
ES grieta de desícación
FR fente de dessiccation

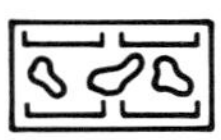

413. FLAT-PEBBLE BRECCIA: a product indicating an advanced step in the sequence desiccation, polygonal cracking and breaking of the crust into individual semilithified pebbles. These, when reworked, form a flat-pebble breccia.

Note: See also Definition 304.

NE platte kiezel breccies
DE —
ES brecha de cantos planos
FR brêche à galets mous; brêche de desiccation

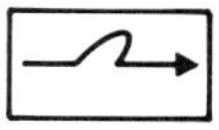

414. INTRASTRATAL FLOW STRUCTURE: a variety of convolute bedding formed by flowage (see Fig. 3–29).

Note: See also Definition 304.

NE vloeistruktuur binnen een laag
DE —
ES estructura de flujo intraformacional
FR lamination convolutée

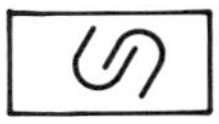

415. INTRASTRATAL CONTORTION (INTRAFORMATIONAL CONTORTION):

Note: See Schlager and Bolz (1977). See also Definition 304.

NE vervorming van/binnen een laag
DE —
ES dislocación intraformacial
FR déformation intra-straté

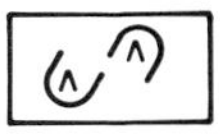

416. ENTEROLITHIC STRUCTURE: a structure resembling the contortions and convolutions of human intestines. The term is applied mainly to anhydrite distortions.

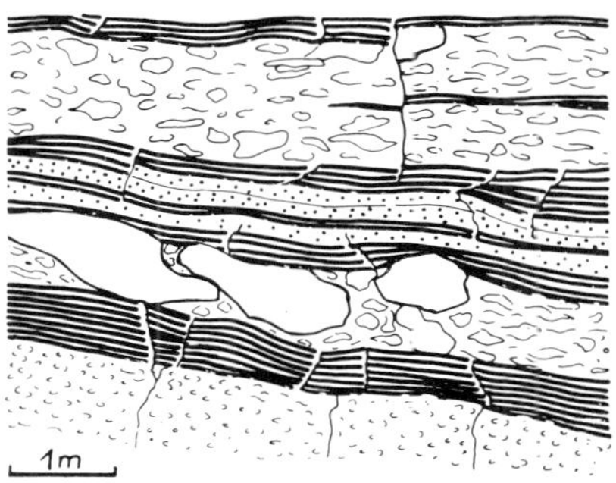

Fig. 3–29. Gypsum breccias interpreted as slumps and mass-flow deposits, with clasts up to 2 m in size, intercalated in laminated gypsum. Quarry Rode, Walhenried, Harz Mountains, Germany, by Schlager and Bolz (1977). By permission of Soc. Econ. Palaeontol. Mineral.

Note
1. See West (1965). Enterolithic structures may give rise to "chicken-wire" structures if contortion ultimately results in irregular evaporite masses in a host rock.
2. See also Definition 304.

NE vervormingen van/binnen een laag; kippengaasstruktuur
DE —
ES estructura enterolítica
FR structure entérolithique

417. SOLUTION BRECCIA: rock with a frame support of large clasts; brecciation and solution fracturing are often due to dissolution of evaporites.

Note: See also Definitions 303, 304 and 323.

NE oplossings breccie
DE Lösungsbrekzie
ES brecha de disolución
FR brèche monogénique de dissolution

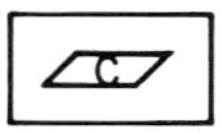

418. DOLOCAST: a cast or impression of a dolomite crystal, preserved in an insoluble residue.

Note: See also Definitions 326 and 351–353.

NE dolomietkristal afdruk
DE —
ES molde de dolomita
FR moulage d'un cristal de dolomite

Mineralogical Modifications

Diagenetic Minerals

Minerals can be another product of diagenesis. They differ from primary minerals in mode, place and time of formation. Their diagenetic origin is determined by the study of their textural, structural and chemical properties. The formation of diagenetic minerals depends on the physical (temperature and pressure) and chemical (nature of interstitial fluids and available ions) environments at the time of their crystallisation. Aspects to be considered are mineralogy, chemical composition, mineral stability, solubility and isotope composition. A knowledge of such aspects enhances understanding of diagenetic processes controlling pore space modifications in carbonates.

Aragonite and Calcite. Many of the dissolution or precipitation problems of diagenetic minerals result from calcium carbonate not behaving as a simple salt in which the equilibrium concentration of both anion and cation is simply the square root of the solubility product. In the calcium carbonate system much of the carbonate entering solution from the solid phase is removed by other reactions, allowing Ca^{2+} concentrations higher then the square root of the solubility product. Also, each of the various crystalline forms of calcium carbonate (i.e., aragonite, high-magnesium calcite and low-magnesium calcite) has its own solubility product. That of aragonite is almost twice that of low-magnesium calcite, and that of skeletal high-magnesium calcite is ~10 times that of low-magnesium calcite. Thus a system that contains at least two solid phases and water cannot have water that is in equilibrium with the system. If the water is at near saturation with respect to aragonite, it will be supersaturated with calcite and will precipitate calcite. Conversely, if the water is near saturation with respect to calcite, it could still dissolve more aragonite. Thus equilibrium is unlikely to be achieved without one of the solid phases either completely dissipating or a steady state being attained in which mineral matter is removed from aragonite crystals and precipitated onto calcite crystals. For this reason practically all aragonite has been dissolved in ancient rocks, or it is converted to another mineral (commonly calcite). Calcite so formed has a low magnesium content, which makes it different from most of the calcite skeletal debris.

Diagenetic calcite is formed either by inversion in the solid state from aragonite or by replacement. Inversion must be accompanied by a change in specific volume; the absence of such a change suggests the recrystallisation process involving solution of aragonite, precipitation of calcite and a net loss of calcium carbonate at the site of replacement.

Diagenetic calcite is a common cementing mineral in limestones and takes many textural shapes (p. 201–202). It could replace dolomite (dedolomitisation).

Dolomite. Dolomite occurs mainly as a secondary mineral (see p. 183). Commonly it replaces calcite or aragonite in carbonate deposits. Often relict textures of the original carbonate sediment are preserved in dolomite crystals. These may help to understand the genesis of carbonate sediments prior to dolomitisation.

The mineral dolomite commonly occurs as euhedral rhombs. The precise chemical composition of the crystal is probably related to that of the dolomitising fluid. Dolomite has a higher density than calcite. If dolomitisation involves only a mole-for-mole replacement of calcium by magnesium (local-source dolomitisation in a closed system) and if compaction effects are negligible, replacing the calcite by dolomite should increase the porosity of a sediment. Apart from being the product of replacement, dolomite might also be precipitated as cement. Thus it occurs as an overgrowth on earlier dolomite crystals (rim cement). Distant-source dolomitisation in an open system results in precipitation of dolomite cements. Continued cementation results in overdolomites.

Gypsum and Anhydrite. Gypsum is the hydrous and anhydrite the anhydrous phase of calcium sulfate. Gypsum is stable at near-surface conditions, anhydrite at depths in which the temperature surpasses 42°C, which coincides with ~800 m.

The precise transition depth depends on the geothermal gradient, the salinity and the partial pressure of the interstitial water. The conversion of anhydrite to gypsum depends on the facility of hydration; relict anhydrite is present locally in outcropping bedded gypsum. Anhydrite can be bedded, pore filling or replacing. Both calcium sulfates are relatively soluble minerals. Solution of sulfate from gypsiferous carbonates forms vugs and crystal moulds.

Other Minerals. Other diagenetic minerals in carbonate rocks are usually less significant. Examples are silicate minerals (quartz, chalcedony and opal), sulfide minerals (pyrite and galene), fluorite, apatite, glauconite and micas.

Solubility of Carbonate and Evaporite Minerals

Solubility data on carbonate and evaporite minerals provide insight into the chemical processes involved in the deposition and alteration of carbonate deposits. Subsurface waters moving in the depth direction in which the solubility of a mineral increases tend to dissolve the mineral. Conversely, those moving in the direction of a mineral's decreasing solubility tend to precipitate. Experiments have led to the following conclusions about the solubility of calcite:

1. It increases with an increase in confining pressure (temperature and CO_2 concentration constant).
2. It increases with an increase in CO_2 content (temperature and confining pressure constant).
3. It decreases with an increase in temperature (CO_2 concentration and confining pressure constant).

Moreover, with respect to saturation the following empirical conclusions have been reached:

1. Mixing of two solutions, both saturated with calcium carbonate, results in an undersaturated solution.
2. Removal of water (e.g., by osmosis) from a solution saturated with calcium carbonate results in a supersaturated solution.

The following also appears to be valid. Descending subsurface waters flow in the direction of increasing temperature, thereby decreasing calcite solubility. This often makes it uncertain whether dissolution of calcite prevails over precipitation in a descending subsurface water of known chemical composition, saturated with calcite. In areas, however, with a normal geothermal gradient (3°C per 100 m) it appears that the temperature effect predominates over the pressure effect. Thus calcite precipitation occurs. Published data on dolomite solubility show wide discrepancies. Deductions based on these data result in different, occasionally conflicting hypotheses on dolomitisation.

Experimental investigations on the solubility of anhydrite suggest that descending waters must eventually precipitate anhydrite, because the effect of increasing temperature, which decreases the solubility of anhydrite, predominates over the effect of increasing pressure, which increases the solubility.

The solubility of gypsum increases with increasing temperature and pressure and high geothermal gradient.

Application of Solubility Data

Experimental and empirical solubility data provide a basis for predicting the solution or precipitation of diagenetic minerals in diagenetic environments with known chemical parameters. Solubility data thus can be used in exploration and production geology to predict or explain certain reservoir rock distribution patterns.

The study of oxygen and carbon isotope ratios in carbonate sedimentary rocks aids in gaining a better understanding of the environment of formation of carbonate minerals. Details are described in most modern handbooks.

Investigations of the mineralogy, chemical composition, mineral stability, solubility and isotope composition of diagenetic minerals all contribute to our knowledge of the diagenetic changes in Recent and ancient carbonates. Such hypotheses, in turn, are reference models that help in predicting and mapping the distribution of diagenetic minerals.

4

Carbonate Reservoirs

4–1 CARBONATE RESERVOIR PARAMETERS

Introduction

This chapter includes a listing of definitions of porosity types and permeability features. As approximately half of the world's hydrocarbons are produced from carbonate reservoirs, a close look at carbonate reservoir sedimentology is warranted. This is especially so because a decrease in discovery rate (Fitzgerald, 1979), dwindling reserves and fluctuating demand for and prices of hydrocarbons have stimulated interest in enhanced hydrocarbon recovery.

Reservoir capacity and production performance are governed by reservoir geometry (external size and shape), boundaries (internal or external seals), internal architecture (i.e., the mosaic of carbonate units with texture-related or independent porosity or permeability) and the nature of inter-/intrareservoir conduit systems.

More so than in the case of siliciclastics, a distinction should be made between the present properties of carbonates as reservoirs and their primary depositional porosity distributions. Carbonate reservoirs can be considered to be composed of a number of carbonate fabric units (see Definition 300) that are the result of a specific diagenetic overprint on a deposit with a particular depositional texture. In most cases it is the diagenetic overprint that produces the actual reservoir, although frequently a good amount of the depositional porosity and permeability remains.

It is possible to classify porosity in carbonates using three different systems: (1) primary versus secondary, (2) homogeneously versus heterogeneously distributed porosity, and (3) texture-related versus texture-independent porosity.

The fundamental difference between primary and secondary porosity types is that the former is present in the sediment from the moment of sedimentation onwards, whereas the latter is diagenetic in origin. Primary variations are controlled by the depositional environment (depositional texture) of the carbonate and the degree of lateral and vertical uniformity of particle size and depositional structures like bedding, lamination and cross-bedding. Mechanisms of secondary porosity formation include fracturing, jointing, leaching and replacement.

Texture-related porosity distribution is governed by the depositional textures and/or the original mineral composition of the grains. Such porosity is thus, to a certain extent, predictable. However, neither the designation of primary or secondary nor that of texture-related or -independent gives an indication of the degree of homogeneity of the pore distribution, which is another factor in predicting regional or local reservoir distribution and character. It is particularly significant to understand a carbonate sequence in these terms when extrapolations are attempted from a limited number of data points and when such data points are variable in nature (e.g., core, log, outcrop section).

Homogeneous pore distribution results in reservoir permeability that is controlled by the dominant size of the textural units forming the carbonate framework. The permeability of a reservoir with homogeneous pore distribution frequently increases proportionally with the bulk porosity. Heterogeneous pore distribution, however, results in reservoir permeability that depends largely on the interrelation of individual pores.

Grain-supported rocks are characterised by grains touching each other. Between the grains, lime mud could be present (packstone) or cement or porosity (grainstone) could be present. In the latter case the porosity is usually homogeneously distributed and is primary. It is also possible to have cementation entirely occluding primary porosity, but leaching—acting simultaneously or later than cementation—could preferentially dissolve (parts of) particles. The result is again a homogeneously distributed porosity, which is then of a secondary nature, however. In both cases the dominant textural unit is of grain size (i.e., larger than 0.02 mm), which potentially gives rise to permeabilities in the range of several darcies.

Mud-supported rocks are characterised by a lime mud matrix in which some grains may or may not occur in a "floating" way. If such grains are selectively leached, a rather heterogeneous pore distribution is the result. The dominant size of the main textural element in mud-supported rocks (i.e., the mud particles) is smaller than 0.02 mm. Because of this low value and because of the lack of interrelation between leached "floating" grains that may provide bigger textural units, the permeabilities are usually not likely to surpass millidarcies.

The foregoing is true for consolidated grain-supported and mud-supported rocks. The unconsolidated mud-supported rocks tend to have a rather homogeneous and high primary porosity distribution, because the interconnected fine interparticular porosity not surpassing 0.02 mm forms the basic pore type. Such carbonates have thus continuous equal reservoir potential. Unconsolidated grain-supported carbonates may also have a homogeneous primary intergranular porosity but this could (partly) be filled by fine lime mud, leaving the grain-supported carbonates with a laterally discontinuous reservoir potential.

In both mud- and grain-supported carbonate types, diagenetic processes may considerably alter the reservoir potential after deposition. They either modify the entire carbonate (e.g., by changing the mineral composition such as at dolomitisation) or act selectively on certain fabric elements. In both cases ultimate reservoir permeability is affected, although to a certain degree it remains dependent on the depositional textural units and it is thus predictable.

A reservoir with heterogeneous pore distribution is commonly characterised by pores of a multigenetic origin. Tectonic forces may give rise to fracture and joint openings. Percolating water may preferentially leach crystals of more soluble minerals. Local-source dolomitisation (Murray, 1960) could form vugs. In limestones where low relative permeability occurs in zones, water may be forced through discreet passages (e.g., joints, burrows, bedding junctions), giving rise to intraformational karst, solution channels and collapse breccias.

Basic pore types consist of a number of characteristic pores in homogeneous and in heterogeneous pore systems. Common porosity-occluding minerals are calcite, dolomite and anhydrite cements and sometimes quartz and pyrite. Moreover, internal sedimentation destroys sometimes basic pore types.

Pressure solution processes give rise to concentrations of insoluble matter in the form of horsetails or stylolites. These are commonly found to be barriers to vertical flow in the reservoir.

The depositional environment largely controls the overall reservoir potential of the sediments; so an understanding of the depositional environment and its lateral consistency aid in predicting the lateral continuity of reservoir beds. This is of prime importance for production geologists and reservoir engineers when deciding on a location for a production or injection well. Reservoir heterogeneity makes lateral extrapolation hazardous. It may adversely affect enhanced recovery performances, because even limited communication between layers can have a pronounced effect on cross-flow and breakthrough sweep efficiency. Knowledge of directional permeability within layers aids in selecting the best arrangement of wells.

Definitions 419–443 (Porosity Types)

$\varnothing$

419. POROSITY: the fraction or percentage of rock or soil occupied by void space, usually expressed as a percentage of bulk volume (see Fig. 4–1).

NE porositeit; poriëngehalte
DE Porosität; Porengehalt
ES porosidad
FR porosité

$\varnothing_B$

420. BULK POROSITY (TOTAL POROSITY): the total volume of void space in a rock, usually expressed as a percentage of bulk volume (see Fig. 4–2).

NE totale porositeit
DE totale Porosität; Gesamtporosität
ES porosidad total
FR porosité totale

421. PORE THROAT: the interconnection between individual pores. It is smaller in diameter than the pores themselves and determines porosity effectiveness.

NE poriën verbinding
DE Porenverbindung
ES interconexíon de poros
FR gorge; étranglement

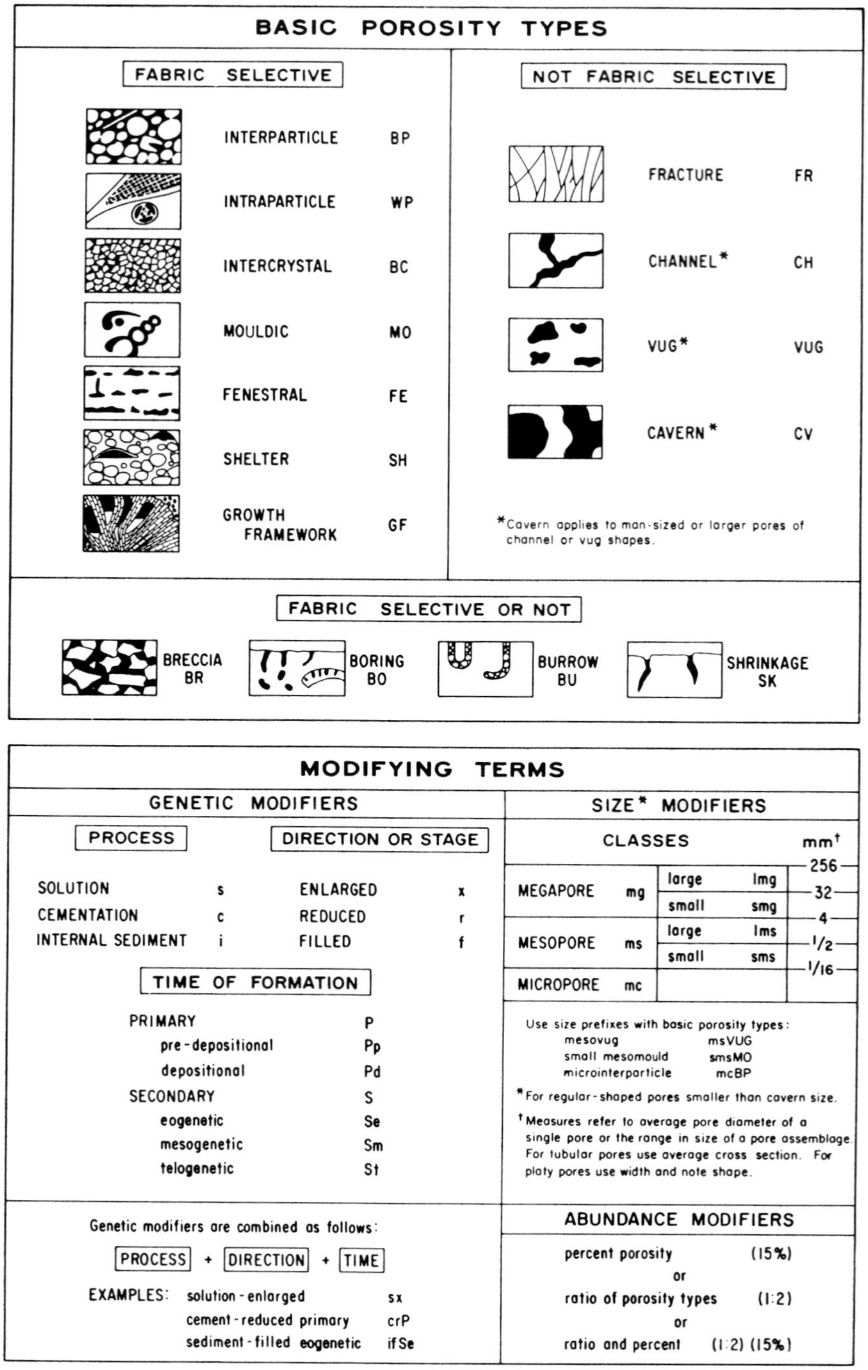

Fig. 4–1. Geological classification of pores in carbonate rocks. Illustration after Choquette and Pray (1970) by permission of the AAPG.

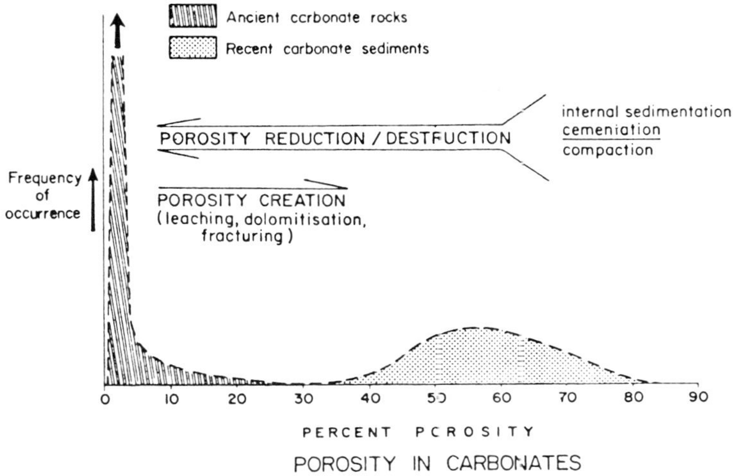

Fig. 4–2. Residual porosity in ancient carbonates is very minor. Reservoirs exist only when the balance of porosity-creating and -occluding processes enhances the residual porosity.

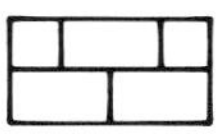

P_1K_1 422. TIGHT: an oil industry term for low permeability. Rock can have high porosity.

NE dicht
DE dicht
ES estanco
FR imperméable; dense

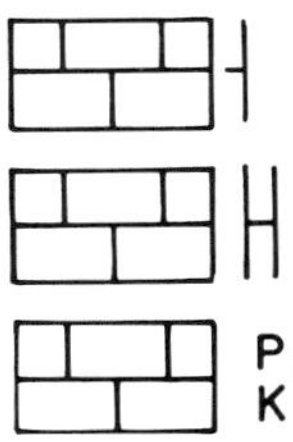

423. EFFECTIVE POROSITY (INTERCONNECTED POROSITY): that fraction of the total porosity through which fluids may flow. The effectiveness is also economically determined, and it depends both on the size of the pore throat and on physical parameters of the fluid, such as wetting characteristics, viscosity, and pressure difference over the porosity.

P_3-7
K_3-7

NE effectieve porositeit; verbonden porositeit
DE nutzbare Porosität
ES porosidad por interconexión; porosidad efectiva
FR porosité efficace

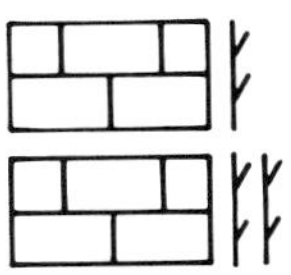

424. INEFFECTIVE POROSITY (DISCONNECTED POROSITY): that fraction of the total porosity through which fluids with certain physical parameters (e.g., viscosity) could not penetrate. It also depends on the size of the pore throat. It includes porosity in isolated vugs.

NE ineffective porositeit; geïsoleerde porositeit
DE nicht nutzbare Porosität; ineffective Porosität
ES porosidad sin interconexión
FR porosité inefficace

425. FRAMEWORK OF POROUS CARBONATE: the frame of granular, particulate or crystalline carbonate in which pores are encased.

NE geraamte van poreus carbonaatgesteente
DE Karbonatgerüst
ES esqueleto de roca carbonatada; armazón de una roca carbonatada porosa
FR squelette de la roche carbonatée

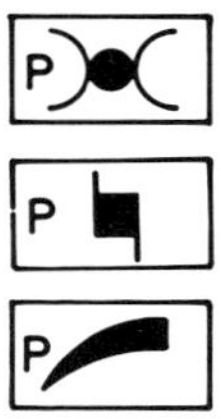

426. PRIMARY POROSITY (SEDIMENTARY POROSITY): the porosity that developed during sedimentation or that was present within sedimentary particles at the time of deposition. Primary porosity includes all predepositional and depositional porosity of a particle, sediment or rock. The primary nature of the carbonate pore types can be indicated by their combination with the pronoun *primary*, for example, primary intergranular or primary intragranular.

NE oorspronkelijke porositeit; sedimentaire porositeit
DE primäre Porosität
ES porosidad primaria; porosidad sedimentaria
FR porosité primaire

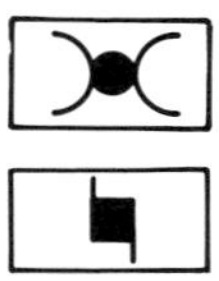

427. INTERGRANULAR POROSITY (INTERPARTICLE POROSITY): porosity between grains or particles.

NE tussen-korrelige porositeit
DE Zwickel-Poren; Interpartikel porosität
ES porosidad intergranular; porosidad entre particulas
FR porosité intergranulaire

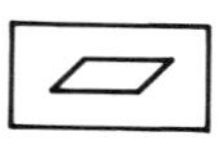

428. INTERCRYSTALLINE POROSITY: porosity between crystals either as a result of a decrease in volume while crystallisation took place from one mineral phase to another, or as a result of incomplete crystallisation, creating a solubility contrast as a result of which the less soluble mineral phase dissolved.

NE interkristallijne porositeit
DE interkristaline Porosität
ES porosidad intercristalina
FR porosité intercristalline

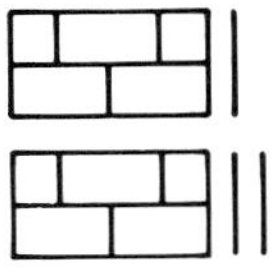

429. MATRIX POROSITY: (1) the porosity of the groundmass of a carbonate rock, in contrast to porosity associated with the particles; (2) the porosity of the massive carbonate rock in contrast to the porosity of the fractures.

NE grondmassaporositeit
DE Matrix Porosität
ES porosidad de la matriz
FR porosité matricielle

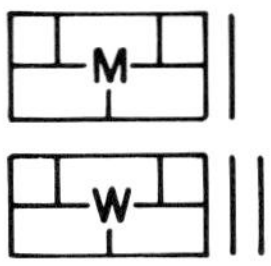

430. MUD-SUPPORTED CARBONATE POROSITY (CHALKY POROSITY): a homogeneous distribution of pores within a mud-supported carbonate. The dominant pore size is usually less than 20 μm.

Note: See also Definition 25.

NE porositeit in een kalkmodder
DE —
ES porosidad ‘‘mud-supported’’
FR porosité dans une roche à grains flottants; porosité ‘‘crayeuse’’

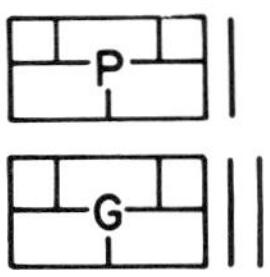

431. GRAIN-SUPPORTED CARBONATE POROSITY: a homogeneous distribution of pores between grains within a grain-supported carbonate rock. The dominant pore size is of grain size; that is, larger than 20 μm.

NE —
DE —
ES porosidad ‘‘grain-supported’’
FR porosité dans une roche à grains jointifs

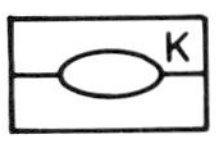

432. KEYSTONE VUG: a primary pore in grainy sedimentary rocks, the size of which is usually larger (millimeters to centimeters) than the size of the grains (units of the sedimentary framework; see Definition 425). Keystone vugs can occur in partially lithified carbonate sands, which have been deposited in high-energy intertidal environments with air bubbles in between the grains. The vugs are the result of these air bubbles.

Note: See also Definition 385.

NE sluitkorrelopening
DE Fenster-Gefüge
ES fenestrae
FR cavités de type ‘‘fenestrae’’; micro-cavernes

433. SHELTER POROSITY: a type of primary interparticle porosity created by the sheltering effect of relatively large sedimentary particles (frequently bioclasts) that partly prevent the infilling of pore space beneath them or in them by finer clastic particles.

Note: See also Definitions 388–390.

NE beschermingsporositeit; honinggraatporositeit
DE Sackporosität; Sackgassenporosität
ES porosidad "shelter"
FR porosité par abri

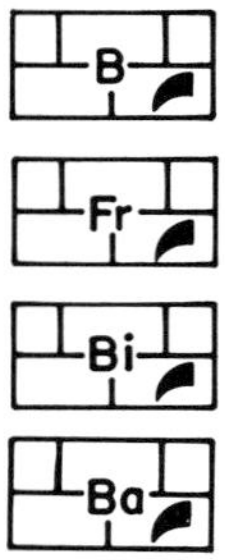

434. FRAMEWORK POROSITY: the primary porosity developed during the in-place growth of a carbonate framework.

NE geraamte porositeit
DE Gerüst-porosität
ES porosidad primaria de crecimiento
FR porosité liée à une charpente organique

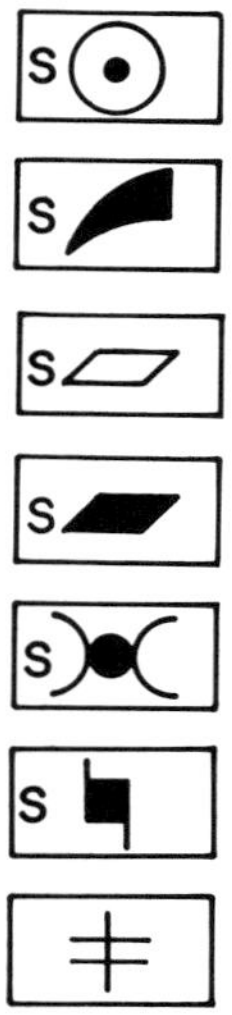

435. SECONDARY POROSITY (DIAGENETIC POROSITY): all porosity developed in a carbonate rock subsequent to its deposition or emplacement, either naturally, through sedimentary, diagenetic or tectonic processes, or artificially, through well stimulation. The secondary nature of carbonate pore types is indicated by the pronoun *secondary*, for example, secondary interparticle, secondary intraparticle, secondary vuggy porosity.

NE secundaire porositeit; gevormde porositeit; diagenetische porositeit
DE sekundäre Porosität
ES porosidad secundaria; porosidad diagenética
FR porosité secondaire

436. TEXTURE-SELECTIVE POROSITY (FABRIC-SELECTIVE POROSITY): secondary porosity that has been created in discrete textural elements of carbonates by diagenetic processes (see Table 4–1).

Table 4–1
Relationship of Texture to Porosity[a]

	Primary	Secondary
Texture related	Intergranular (>20 μm) Intergranular Fine interparticle (<20 μm) Shelter Framework	Intergranular Intraskeletal Vug Mouldic
Not texture related	Matrix Pinpoint	Intragranular Fracture Matrix Pinpoint

[a] Porosity-related terms: no porosity, tight; effective porosity, interconnected porosity; uneffective porosity, disconnected porosity.

Note: See also Definitions 427 and 438.

NE maakselporositeit
DE Gefügeporosität
ES porosidad relacionada con la textura
FR porosité dépendant de la texture de la roche

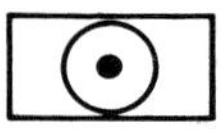

437. INTRAGRANULAR POROSITY (INTRASKELETAL POROSITY): secondary porosity within individual grains or skeletal parts.

NE inkorrelige porositeit
DE Hohlformporosität; Partikel-Porosität; Intra-Partikel-Porosität
ES porosidad intragranular
FR porosité intragranulaire

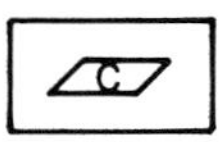

438. MOULDIC POROSITY: texture-related secondary porosity, formed by the selective removal, usually by dissolution, of a former individual constituent of the sedimentary texture.

Note: See also Definition 301.

NE afdrukporositeit
DE Partikellösungs-Porosität
ES porosidad por molde; porosidad debida a moldes
FR porosité vacuolaire de vides de dissolution

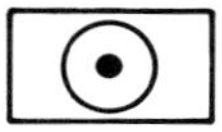

439. INTRAOOID POROSITY: secondary, texture-related porosity formed by dissolution or leaching in the lamellae of ooids. A specific form of intragranular porosity.

Note: See also Definition 301.

NE porositeit in opgeloste ooïde-lamellen
DE Lösungsporosität in Ooiden
ES porosidad intraooidal; porosidad intraoolítica
FR porosité intra-oolithique

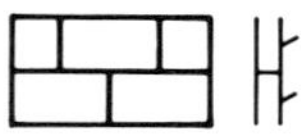

440. PINPOINT POROSITY: a term for very small secondary pores in carbonate rocks, usually of the mouldic or intercrystalline porosity types.

Note: See also Definition 438.

NE speldeprik porositeit
DE —
ES porosidad "pinpoint"
FR microporosité en trous d'épingle

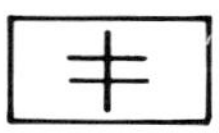

441. TEXTURE-INDEPENDENT POROSITY: porosity that is not selectively present in particular depositional textural elements (e.g., secondary solution vugs).

NE niet maaksel-gebonden porositeit
DE —
ES porosidad no relacionada con la textura
FR porosité independante de la texture

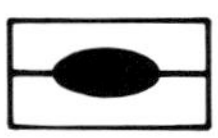

442. VUG POROSITY: porosity which is not fabric selective; often solution-enlarged moulds, sometimes solution channels; generally larger than grains.

Note: See also Definition 303.

NE —
DE Gesteinslösungs-Porosität
ES porosidad debida a cavidades; porosidad vacuolar
FR porosité vacuolaire

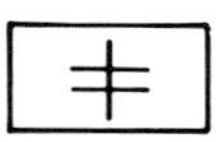

443. FRACTURE POROSITY: secondary porosity formed by fractures; not texture related.

NE breukporositeit
DE Kluftporosität

ES porosidad de fractura
FR porosité de fracture

4–2 CARBONATE RESERVOIRS: EXPLORATION APPROACHES

Carbonate reservoirs are the result of depositional and diagenetic processes. One must take both into account when trying to estimate porosity distribution in the subsurface. Attempts to predict carbonate reservoir distribution meet with variable success for different types of reservoirs. In the following paragraphs attention will be focussed on questions such as What are the different types of carbonate reservoirs? and What parameters should be studied and mapped? in order to arrive at a genetic reservoir classification that aids attempts to predict their occurrence.

Genetic Reservoir Types

Two aspects are significant in the description of carbonate reservoirs and establishing genetic types of reservoirs: (1) the mode of deposition and (2) the origin of pore space.

Mode of Deposition

Carbonate rocks acquire different compositional and textural characteristics during deposition (see pp. 70–71). A sequence of carbonates will therefore consist of many rock types.

Variations are apparent in composition (grain types), depositional structures and textures. If such variations, however, are different manifestations of one single sedimentation process or of a group of closely interrelated processes, then the carbonate rocks concerned could be considered to constitute a facies unit (pp. 78–80). Properly interpretated, facies can then become mappable genetic units that are predictable. Consequently, reservoir characteristics, typical of such facies, are also predictable.

Origin of Pore Space

Pore type and permeability in carbonates are to a certain extent related to depositional textures and diagenetic fabrics. Pore types can be distinguished on the basis of textural criteria (e.g., interparticle or intercrystalline). Thus a number of descriptive basic pore types can be recognised. An interpretative approach is to group the various reservoirs genetically by referring to the process which is directly or indirectly responsible for the origin of the predominant pore type. Thus one recognises

1. Residual primary porosity (pore space acquired during sedimentation and partially preserved)
2. Dissolution porosity (pore space acquired during early or late diagenetic solution)
3. Porosity created as a result of volume changes during mineral replacements (e.g., dolomitisation)

4. Diagenetic fracture porosity (fracturing during diagenesis or as a result of diagenesis)
5. Tectonic fracture porosity (fracturing during tectonic deformation)
6. Composite porosity (combination of two or more processes listed above)

Recognition of these pore-creating processes and their ensuing basic pore types is important because it is the first step towards proper recognition of the process creating the reservoir, and it aids in predicting carbonate porosity trends.

Descriptive Reservoir Types

Pores can be related or unrelated to the depositional texture of (diagenetically altered) carbonates. This fact forms the basis of descriptive classifications. Pores are also of different structure. Pore structure encompasses pore shape and size and is qualitatively indicated by the relation between porosity and permeability. Low porosity accompanied by high permeability indicates larger pores and, generally, low connate water content because of the low capillary pull. High porosity accompanied by low permeability suggests smaller pores and, generally, higher connate water content.

A descriptive classification places emphasis on the flow of fluids (permeability and relative permeability), fluid distribution (capillary pressure) and flow of electricity (resistivity), rather than on the mineral content or any genetic criteria (as already discussed). A useful descriptive classification of carbonates is based upon the following:

1. The texture of the matrix (crystal size; nature of crystal boundaries). Three classes are significant: a completely crystalline matrix, a chalky matrix and a particulate or saccharoidal matrix.
2. The character of the visible pore structure. It is practical to recognise three pore-size classes using a low-powered microscope or hand lens: (a) no visible pores, (b) pores between 0.1 and 0.01 mm, and (c) pores surpassing 0.1 mm in diameter.

Recognition of the three textural classes and three pore-size classes allows a tentative petrophysical evaluation in terms of potential productivity of certain strata. This approach, however, is of limited use for extrapolation of porosity trends. To this aim attention has to be focussed on the genetic features of the carbonate reservoir.

Subdivision of Carbonate Reservoirs into Genetic Units

Genetic units within reservoirs can be based on the following criteria:

1. Interpretation of facies units in environmental terms and extrapolation of facies belts into untested areas using standard concepts such as progradation and transgression
2. Description of diagenetic processes, interpretation of the resulting diagenetic facies in terms or depth and/or time zones and extrapolation of diagenetic facies trends into untested areas by extending recognised diagenetic gradients

Sedimentary intervals that can be characterised in terms of these criteria are here referred to as carbonate fabric units. In general, certain types of porosity can be expected in particular types of carbonates, and if they occur in specific carbonate fabric units, they become predictable to a certain extent.

Shallow Marine Carbonates

Residual primary porosity can be expected in organic boundstone or packstone masses or in accumulations of lime sand (grainstones). Lagoonal or tidal flat carbonates and micritic sediments have low reservoir potential unless they have been altered by solution, replacement, fracturing or a combination of these processes. Their initial high porosity is very rapidly occluded by early compaction and cementation.

Deeper Marine Carbonates

Carbonates deposited in deeper marine environments retain residual primary porosity with greater ease in fine-grained sediments. Coccolith oozes have a stable calcitic mineralogy. After induration of these oozes into chalk, this chalk is less likely to be altered by solution by meteoric water or by replacement processes than the aragonite and high-magnesium calcitic micritic sediments in lagoons and tidal flats carbonates. Some deep marine chalk deposits have enhanced permeability due to fracturing.

Carbonate masses that have moved into the deeper marine realm (turbidites, slumps, etc.) often offer good promise of reservoir development because the reservoir rock is of shallow marine provenance and frequently made up of minerals that are unstable in the deeper marine depositional environment due to carbonate dissolution with depth. Another reason is that transportation after initial deposition enhanced porosity and permeability, as has become apparent from work in North Sea chalks.

Prediction of Carbonate Reservoirs

Indirect prediction of carbonate reservoirs depends on mapping of sedimentary and/or diagenetic facies. Genetic grouping of carbonate reservoirs reduces the number of maps needed. An attempt should be made to map the distribution and abundance of a sedimentary or diagenetic product responsible for the following:

1. The origin of pores space. Thus maps of sedimentary facies distribution allow extrapolation of the distribution of residual primary porosity; maps of diagenetic facies distribution allow the prediction of distribution of secondary porosity.
2. The preservation or destruction of pore space. A map of sedimentary facies and an overlay of such diagenetic features as cementation or compaction aid in the prediction of distribution of residual primary porosity.
3. The creation of pore space. A map of sedimentary facies could be overlaid by a map of such diagenetic features as leaching and dissolution. This might show trends of porosity creation. The relation between tectonics and late diagenesis could lead to the construction of a map of fracture patterns pre-

dicting the distribution of solution or replacement porosity. A map showing spatial distribution of fractures could aid in predicting tectonic fractures.

Such mapping exercises and the extrapolation that is required demand a good number of data points in which logs, sidewall samples or cores are available. Calibration of facies recognised in cores or sidewall samples—with their characteristic log response (in particular, on the sonic and the compensated neutron-density logs)—will form a basis for regional extrapolation, even into areas in which no core material is available.

4–3 PROCEDURAL GUIDELINES

Where stratigraphic elements play an important role in trapping, sophisticated carbonate reservoir studies utilising all available data can help to delineate the extent of porosity. The work involves, first of all, recognition of the reservoir type. Three different types of reservoir are possible: (1) those with facies-related preserved primary porosity, (2) those with facies-related secondary porosity and (3) those with no facies-related porosity. If reservoirs have facies-related preserved primary porosity, these should be delineated as follows:

1. Such carbonate facies as reefs, calcarenites, etc., that contain primary porosity should be mapped.
2. The nature and distribution of such diagenetic processes as cementation and compaction that have destroyed primary porosity should be identified.
3. On the basis of reservoir beds delineated in this way, some extrapolations can be attempted with the use of wireline logs by mapping the distribution of depositional facies on the basis of existing models.

If reservoir beds have facies-related preserved secondary porosity, one might follow a procedure similar to that described in order to delineate the distribution of such a reservoir.

If reservoir beds do not have facies-related preserved porosity, reservoir bed localisation is very difficult. An attempt can then be made to

1. Delineate the porous zones at all data points available
2. Identify the nature of the diagenetic porosity-creating or -occluding process
3. Map the products of such processes (e.g., calcite cement and dolomite and distribution of vugs or fractures) in all available data points
4. Extrapolate the mapped products of such diagenetic processes along trends using criteria such as "intensity" of diagenesis and log character of zones characterised by the recognised diagenetic process.

From the foregoing it is clear that exploration in carbonate areas where available information is incomplete requires a comprehensive knowledge of the depositional, diagenetic and structural histories of the objective formations on a regional scale. Delimiting possible reservoir beds requires at least some well and/or outcrop information in order to

1. Define porous zones
2. Map carbonate facies within these zones
3. Recognise the porosity-modifying diagenetic processes

4. Map the products of such diagenetic processes
5. Superimpose maps of carbonate depositional and diagenetic facies, showing the possible distribution of primary and secondary porosity, respectively, and maps of cementation and compaction showing where the preservation of these porosities might be expected

Such a procedure could result in a broad delineation of the distribution of porous and permeable carbonate trends. On this basis, extrapolations of depositional and diagenetic facies could be attempted. Such extrapolations would be based on existing theoretical depositional and diagenetic models and observed trends.

Of course, the search for reservoirs is only part of the game. Considerations of reservoir distribution should be complemented by those of trapping, occurrences of source beds and the migration history of hydrocarbons. However, an early understanding of the types of carbonate reservoirs that can be expected could aid in deciding whether subtle stratigraphic/diagenetic traps could occur together with more conventional structural traps.

4–4 ROLE OF THE COMPUTER IN CARBONATE RESERVOIR PREDICTION

The work involved in predicting the location of carbonate reservoirs includes correlation, mapping of sedimentary facies, mapping of diagenetic facies and recognition of the origin of pore spaces.

Correlation

Correlation can be established on the basis of stratigraphic palaeontology, lithological properties or geophysical logs. For these three types of correlations, data points are needed. Recently a fourth possibility has been developed; seismic expression of big units leads to basin-wide analyses of sequences of carbonates within the frame of seismostratigraphy. Further definition of the reliability, boundaries and possibilities of seismostratigraphy in carbonates is called for in view of the limited density contrasts and consequently the limited resolution that usually occur between various carbonate units. A fruitful connection between carbonate reservoir detection and correlation might be made with computer-compatible stored geological information, allowing for statistical evaluation of the geological input.

Valid correlation is necessary in order to construct meaningful maps showing the distribution of carbonate facies and to delineate useful genetic units. Efforts to improve correlation should include improved definition of geophysical and lithological markers with local and/or regional time significance. Correlation with the aid of geophysical data (both logs and seismic sections) involves the correlation of physical properties of sedimentary rocks which cannot be carried out successfully unless one has a thorough knowledge of carbonate sedimentation and diagenetic principles.

Mapping Sedimentary Facies

Sedimentary facies are defined on the basis of the sediment composition, depositional texture and sedimentary structure. These physical properties are mappable. They can also be filed in the computer and retrieved in endless combinations, allowing construction of a host of trend maps.

Facies distribution patterns are the spatial reflection of genetic units; sedimentary cycles, the temporal ones. Genetic units are recognised following a sound interpretation of sediment composition, textures and structures in terms of sedimentary processes. Comparison with Recent sediments allows a correlation with properties and processes that can now be observed. Mapping sedimentary facies and understanding their temporal and spatial variations permits the prediction of one facies on the basis of the known distribution of other facies.

Mapping Diagenetic Facies

Diagenetic facies are recognised and defined in terms of active diagenetic processes and products formed. Serious efforts have been made to separate these two aspects of diagenesis and thus classify the rock history leading to reservoir formation. Recent diagenesis and ancient diagenesis have been studied in order to understand diagenetic features in terms of the processes that produce them and the products that characterise certain depth zones. Emphasis has been placed on understanding the causes and effects of cementation, solution and dolomitisation. Some reference models of dolomitisation and cementation are available, and a generalised space–time diagram, in which sequences of diagenetic events can be plotted, has been designed (Fig. 3–6). It could also serve as a basis for computer-compatible data storage. Reference models may aid in lateral extrapolation; the generalised space–time diagram encourages a uniform and mutually comparable approach, and the computer-compatible data storage makes available basic information that can be manipulated in such a way that new approaches can be defined.

Origin of Pore Space

An interpretation of the origin of pore space requires a detailed understanding of the depositional or diagenetic textures and an appreciation of grain types, mineralogy, sedimentary structures and diagenetic structures. The prediction of the distribution of porosity is limited to the understanding of the origin of the various genetic reservoir types. Several combinations of depositional and/or diagenetic processes can lead to the preservation or creation of pore space in carbonates. Genetic grouping of reservoir types focusses attention on possible reservoir types and the parameters that have to be selected to predict their location. Case studies focus attention on actually identified reservoir types and provide feedback for the genetic grouping system.

4–5 COMPUTER-COMPATIBLE CARBONATE DATA STORAGE

A computer-compatible input system for carbonate sedimentological data is of considerable help in the systematic recording of the depositional and diagenetic carbonate compositional, textural and structural elements that collectively determine reservoir characteristics. A prerequisite for such a data-handling system is the use of a systematic and hierarchical arrangement of basic concepts, definitions and terms such as that outlined in the previous chapters.

For computer processing the terms require a standardised format which is defined by a numerical or alpha numerical code with a certain number of digits or characters. Some codes have already been defined in Shell's Standard Legend. Where possible, these have been adopted in the present work.

The general description of sedimentological data for storage purposes should be divided into a depositional and a diagenetic part.

Storage of Depositional Data

Six major aspects are recognisable in descriptions of depositional data:

1. Lithological information. In carbonates this pertains mainly to mineralogy, although provisions should be made in any file system to file data on impure lithotypes (Sections 1–1 and 1–2).*
2. The particulate composition. Here, the nature of particles can be filed (Section 1–3).
3. Sedimentary structures and bedding characteristics should be filed separately (Section 1–4).
4. Depositional textures and some information on grain size, sorting and roundness can be stored when relevant (Section 1–4).
5. Environment of deposition. Various aspects of such environments can be stored (Chapter 2).
6. Porosity information (Chapter 4).

Storage of Diagenetic Data

As part of reservoir studies, diagenetic investigations aim at the recognition of porosity-creating and -occluding processes in a time and/or depth setting. Diagenetic processes are ongoing events both in time and in space, leaving products that reflect various states of this process. The processes may affect specific particles, minerals or the entire rock. The products of diagenetic processes can broadly be subdivided into one or more of the following categories:

1. Diagenetic minerals (Section 3–4).
2. Diagenetic textures (Section 3–4).

Here, a host of descriptive terms related to crystals size, shape and boundaries can be found:

1. Diagenetic structures (Section 3–4).
2. Genetic pore types (Section 4–1).

Consequently any input system should make provisions to file successively

1. The type of diagenetic process (Section 3–3)
2. The rock material that is affected
3. The results (at least two possibilities) of the activity of the process
4. Details on the ensuing porosity (e.g., origin and the type of the pore (Section 4–1)
5. The time and/or depth setting of both processes and products

Provisions should be made to repeat the filing procedure outlined above several times. Only then can diagenetic sequences (simple as well as complex ones) be described in great detail.

* Section numbers refer to Chapters in the present book.

References

Adams, J. E., and Rhodes, M. L. (1960). Dolomitization by seepage refluxion. *Am. Assoc. Pet. Geol. Bull.* **44,** 1912–1921.

Ahr, W. M. (1973). The carbonate ramp: An alternative to the shelf model. *Trans.–Gulf Coast Assoc. Geol. Soc.* **23,** 221–225.

Aitken, J. D. (1967). Classification and environmental significance of cryptalgal limestones and dolomites, with illustrations from the Cambrian and Ordovician of Southwestern Alberta. *J. Sediment. Petrol.* **37,** 1163–1178.

Alexandersson, T. (1973). Intragranular growth of marine aragonite and Mg-calcite: Evidence of precipitation from supersaturated sea water. *J. Sediment. Petrol.* **43,** 441–460.

Alexandersson, T. (1974). Carbonate cementation in coralline algal nodules in the Skagerrak, North Sea: Biochemical precipitation in undersaturated waters. *J. Sediment. Petrol.* **44,** 7–26.

Assereto, R. L. M., and Kendall, C. G. St. C. (1977). Nature, origin and classification of peritidal tepee structures and related breccias. *Sedimentology* **24,** 153–210.

Badiozomani, K., Mackenzie, F. T., and Thorstenson, D. C. (1977). Experimental carbonate cementation: Salinity, temperature and vadose-phreatic effects. *J. Sediment. Petrol.* **47,** 529–542.

Baker, P. G., and Kastner, M. (1981). Constraints on the formation of sedimentary dolomite. *Science* **213,** 213–216.

Bartok, P., Reijers, T. J. A., and Juhasz, I. (1981). Lower Cretaceous Cogollo Group, Maracaibo basin, Venezuela: Sedimentology, diagenesis and petrophysics. *Am. Assoc. Pet. Geol. Bull.* **65,** 1110–1134.

Basan, P. B., ed. (1978). Trace fossil concepts. Soc. Econ. Paleontol. Mineral. Short Course 5.

Bathurst, R. G. C. (1971). Carbonate sediments and their diagenesis. *Dev. Sedimentol.* **12.**

Bricker, O. P., ed. (1971). "Carbonate Cements," Stud. Geol., No. 19. Johns Hopkins Univ. Press, Baltimore, Maryland.

Burchette, T. P. (1981). European Devonian reefs: A review of current concepts and models. *Spec. Publ.—Soc. Econ. Paleontol. Mineral.* **30,** 85–142.

Calvet Rovira, F., Pomar Gomá, L., and Esteban Cerdá, M. (1975). Las rizocreciones del Pleistoceno de Mallorca. *Inst. Invest. Geol. Univ. Barcelona* **30,** 35–60.

Carozzi, A. V. (1960). "Microscopic Sedimentary Petrography." Wiley, New York.

Carrison, R. E., and Kennedy, W. J. (1977). Origin of solution seams and flaser structures in Upper Cretaceous chalks of Southern England. *Sediment. Geol.* **19,** 107–137.

Cayeux, L. (1970). "Sedimentary Rocks of France. Carbonate Rocks (Limestone and Dolomites)" (Translated and updated by A. V. Carozzi). Hafner, Darien.

Chilingar, C. V., Bissell, H. J., and Fairbridge, R. W., eds. (1976). "Carbonate Rocks," Dev. Sedimentol., Vols. 9A and 9B. Elsevier, Amsterdam.

Choquette, P. W., and Pray, R. C. (1970). Geological nomenclature and classification of porosity in sedimentary carbonates. *Am. Assoc. Pet. Geol. Bull.* **54,** 207–250.

Cloud, P. E., Jr. (1962). Environment of calcium carbonate deposition west of Andros Island, Bahamas. *Geol. Surv. Prof. Pap. (U.S.)* **350,** 1–138.

Cook, H. E., and Enos, P., eds. (1977). "Deep-water Carbonate Environments," Spec. Publ., No. 25. Soc. Econ. Paleontol. Mineral., Tulsa, Oklahoma.

Cook, H. E., and Mullins, H. T. (1983). Basin Margin Environment. *In* "Carbonate Depositional Environments," (P. A. Scholle, D. G. Bebout, and C. H. More, eds.) *Am. Assoc. Pet. Geol. Memo* **33,** 539–618.

Cook, H. E., Hine, A. C., and Mullins, H. T., eds. (1983). Platform margin and deep water carbonates. Soc. Econ. Paleontol. Mineral. Short Course 12.

Corley, C. D. (1977). Origin of distorted ooliths and pisoliths. *J. Sediment. Petrol.* **47,** 554–564.

Crevello, P. D., and Harris, P. M., eds. (1985). Deep water carbonates: a core workshop. Soc. Econ. Paleontol. Mineral. Core Workshop 6.

Dahanayake, K. (1977). Classification of oncoids from the Upper Jurassic Carbonates of the French Jura. *Sediment. Geol.* **18,** 337–353.

Daly, R. N. (1915). The glacial-control theory of coral reefs. *Proc. Am. Acad. Arts Sci.* **51,** 157–251.

Darwin, C. (1874). "The Structure and Distribution of Coral Reefs," 2nd ed. Smith, Elder, London.

Deelman, J. C. (1978). Protodolomite redefined; discussion. *J. Sediment. Petrol.* **48,** 1004–1007.

Dunham, R. J. (1962). Classification of carbonate rocks according to depositional texture. *In* "Classification of Carbonate Rocks" (W. E. Ham, ed.), Mem. No. 1, pp. 108–121. Assoc. Pet. Geol., Tulsa, Oklahoma.

Dunham, R. J. (1969a). Early vadose silt in Townsend Mount (reef), New Mexico. *Spec. Publ.—Soc. Econ. Paleontol. Mineral.* **14,** 139–181.

Dunham, R. J. (1969b). Vadose pisolite in the Capitan reef (Permian), New Mexico and Texas. *Spec. Publ.—Soc. Econ. Paleontol. Mineral.* **14,** 182–191.

Dunham, R. J. (1970). Stratigraphic reefs versus ecologic reefs. *Am. Assoc. Pet. Geol. Bull.* **54,** 1931–1932.

Dunoyer de Segonzac, G. (1968). The birth and development of the concept of diagenesis (1866–1966). *Earth-Sci. Rev.* **4,** 153–201.

Einsele, G., and Seilacher, A., eds. (1982). "Cyclic and Event Stratification." Springer-Verlag, Berlin and New York.

Elf Aquitaine (1975a). "Essai de caractérisation sédimentologique des dépôts carbonatées. 1. Eléments d'analyse." Elf Aquitaine Centre de Recherches, Boussens et Pau.

Elf Aquitaine (1975b). "Essai de caractérisation sédimentologique des dépôts carbonatés. 2. Eléments d'interprétation." Elf Aquitaine Centre de Recherches, Boussens et Pau.

Ellis, P. M., Crevello, P. D., and Elink, L. S. (1985). Upper Jurassic and lower Cretaceous deep-water buildups. Abenaki formation, Nova Scotia Shelf. *In:* "Deep Water Carbonates," (P. D. Crevello and P. M. Harris, eds.) Soc. Econ. Paleontol. Mineral. Core Workshop 6.

Embry, A. F., and Klovan, E. J. (1972). Absolute water depths limits of Late Devonian paleoecological zones. *Geol. Rundsch.* **61**(2).

Esteban, C. M. (1976). Vadose pistolite and caliche. *Am. Assoc. Pet. Geol. Bull.* **60,** 2048–2057.

Evamy, B. D. (1963). The application of a chemical staining technique to a study of dedolomitisation. *Sedimentology* **2,** 164–170.

Fisher, J. H., ed. (1977). "Reefs and Evaporites-Concepts and Depositional Models," Stud. Geol., No. 5. Am. Assoc. Pet. Geol., Tulsa, Oklahoma.

Fitzgerald, T. A. (1980). Giant field discoveries 1968–1978: An overview. *In* "Giant Oil and Gas Fields of the Decade 1968–1978," (M. T. Halbouty, ed.), Mem. No. 30, pp. 1–5. Am. Assoc. Pet. Geol., Tulsa, Oklahoma.

Flügel, E. (1978). "Mikrofazielle Untersuchungsmethoden von Kalken." Springer-Verlag, Berlin and New York.

Flügel, E. (1982). "Microfacies Analysis of Limestones." Springer-Verlag, Berlin and New York.

Folk, R. L. (1959). Practical petrographic classification of limestones. *Am. Assoc. Pet. Geol. Bull.* **43,** 1–38.

Folk, R. L. (1962). Spectral subdivision of limestone types. *In* "Classification of Carbonate Rocks" (W. E. Ham, ed.), Mem. No. 1, pp. 62–84. Am. Assoc. Pet. Geol., Tulsa, Oklahoma.

Folk, R. L. (1965). Some aspects of recrystallisation in ancient limestones. *Spec. Publ.—Soc. Econ. Paleontol. Mineral.* **13,** 14–49.

Folk, R. L., and Land, L. S. (1975). Mg/Ca Ratio and Salinity: Two controls over crystallisation of dolomite. *Am. Assoc. Pet. Geol. Bull.* **59,** 60–68.

Friedman, G. M. (1959). Identification of carbonate minerals by staining methods. *J. Sediment. Petrol.* **29,** 87–97.

Friedman, G. M. (1964). Early diagenesis and lithification in carbonate sediments. *J. Sediment. Petrol.* **34,** 777–813.

Friedman, G. M., and Sanders, J. E., eds. (1978). "Principles of Sedimentology." Wiley, New York.

Füchtbauer, H. (1974). Sediments and sedimentary rocks. I. *In* "Sedimentary Petrology" (W. V. Engelhardt, H. Füchtbauer, and G. Müller, eds.), Part II.

Füchtbauer, H., and Müller, G. (1970). Sedimente und Sedimentgesteine. *In* "Sediment-Petrologie 2." Schweizenbart, Stuttgart.

Gaines, A. M. (1977). Protodolomite redefined. *J. Sediment. Petrol.* **47,** 543–564.

Gaines, A. M. (1978). Protodolomite redefined; reply. *J. Sediment. Petrol.* **48,** 1009–1011.

Gidman, J. (1978). Protodolomite redefined; discussion. *J. Sediment. Petrol.* **48,** 2007–2008.

Ginsburg, R. N. (1953). Intertidal erosion on the Florida Keys. *Bull. Mar. Sci. Gulf Carrib.* **3,** 55–69.

Ginsburg, R. N., ed. (1975). "Tidal Deposits. A Casebook of Recent Examples and Fossil Counterparts." Springer-Verlag, Berlin and New York.

Ginsburg, R. N., and James, N. P. (1974). Holocene carbonate sediments of continental shelves. *In* "The Geology of Continental Margins" (C. A. Burk and C. L. Drake, eds.), pp. 137–155. Springer-Verlag, Berlin and New York.

Ginsburg, R. N., Bricker, O. P., Wanless, H. L., and Garrett, P. (1970). Exposure index and sedimentary structures of a Bahama Tidal flat. *Geol. Soc. Am. Abstr. Prog. Annu. Meet.* pp. 744–745.

Ginsburg, R. N., Schroeder, J. H., and Shinn, E. A. (1971). Recent synsedimentary cementation in subtidal Bermuda reefs. *In* "Carbonate Cements" (O. P. Bricker, ed.), pp. 54–56. Johns Hopkins Univ. Press, Baltimore, Maryland.

Grabau, A. W. (1913). "Principles of Stratigraphy." A. G. Seiler & Co., New York.

Halbouty, M. T., ed. (1980). "Giant Oil and Gas Fields of the Decade: 1968–1978," Mem. No. 30. Am. Assoc. Pet. Geol., Tulsa, Oklahoma.

Ham, W. E., ed. (1962). "Classification of Carbonate Rocks," Mem. No. 1. Am Assoc. Pet. Geol., Tulsa, Oklahoma.

Hardie, L. A., ed. (1977). "Sedimentation of the Modern Tidal Flats of Northwest Andros Island, Bahamas," Stud. Geol., No. 22. Johns Hopkins Univ. Press, Baltimore, Maryland.

Harrison, R. S. (1977). Caliche profile: Indicators of near-surface subaerial diagenesis, Barbados, West Indies. *Bull. Can. Pet. Geol.* **25**(1), 125–173.

Heckel, P. H. (1972). Recognition of ancient shallow marine environments. *Spec. Publ.—Soc. Econ. Paleontol. Mineral.* **16,** 226–286.

Heckel, P. H. (1974). Carbonate buildups in the geologic record: A review. *Spec. Publ.—Soc. Econ. Paleontol. Mineral.* **18,** 90–154.

Hedgpeth, J. W. (1957). Classification of marine environments. *Mem.—Geol. Soc. Am.* **67,** 17–27.

Hopley, D. (1982). "The Geomorphology of the Great Barrier Reef." Wiley, New York.

Horodyski, R. J., and Bloeser, B. (1977). Laminated algal mats from a coastal lagoon, mormona, Baja California, Mexico. *J. Sediment. Petrol.* **47,** 680–698.

Horowitz, A. S., and Potter, P. E. (1971). "Introductory Petrography of Fossils." Springer-Verlag, Berlin and New York.

Hsü, K. J., and Jenkyns, H. C., eds. (1974). "Pelagic Sediments: On Land and Under Sea," Spec. Publ. Int. Assoc. Sedimentol., Vol. 1. Blackwell, Oxford.

Hsü, K. J., and Siegenthaler, C. (1969). Preliminary experiments and hydrodynamic movement induced by evaporation and their bearing on the dolomite problem. *Sedimentology* **12,** 11–25.

Irwin, M. L. (1965). General theory of epeiric clear water sedimentation. *Am. Assoc. Pet. Geol. Bull.* **49,** 445–459.

James, H. P., and Ginsburg, R. M. (1979). The seaward margin of Belize Barrier and Atoll reefs. *Spec. Publ.—Int. Assoc. Sedimentol.* **3,** 1–191.

Kaasschieter, J. P. H., and Reijers, T. J. A., eds. (1983). "Petroleum Geology of the Southeastern North Sea and the Adjacent Onshore Areas," Spec. issue. Geologie & Mijnbouw, The Hague.

Kendall, A. C. (1977). Origin of dolomite mottling in Ordovician limestones from Saskatchewan and Manitoba. *Bull. Can. Pet. Geol.* **25**(3).

Kendall, A. C., and Broughton, P. L. (1978). Origin of fabrics in speleothems composed of columnar calcite crystals. *J. Sediment. Petrol.* **48,** 519–538.

Knox, G. J. (1977). Calcite profile formation, Saldanka Bay (South Africa). *Sedimentology* **24,** 657–674.

Kobluk, D. R., and Risk, M. J. (1977). Calcification of exposed filaments of endolithic algae, micrite envelope formation and sediment production. *J. Sediment. Petrol.* **47,** 517–528.

Land, L. S. (1973a). Contemporaneous dolomitization of Middle Pleistocene reefs by meteoric water, North Jamaica. *Bull. Mar. Sci.* **23,** 64–92.

Land, L. S. (1973b). Holocene meteoric dolomitization of Pleistocene limestones. North Jamaica. *Sedimentology* **20,** 411–424.

Laporte, L. F., ed. (1974). "Reefs in Time and Space. Selected Examples from the Recent and Ancient," Spec. Publ., No. 18. Soc. Econ. Paleontol. Mineral., Tulsa, Oklahoma.

Larsen, G., and Chilingar, G. V., eds. (1967). "Diagensis in Sediments," Dev. Sedimentol., No. 8. Elsevier, Amsterdam.

Leighton, W. M., and Pendexter, C. (1962). Carbonate rock types. In "Classification of Carbonate Rocks" (W. E. Ham, ed.), Mem. No. 1, pp. 33–61. Am. Assoc. Pet. Geol., Tulsa, Oklahoma.

Logan, B. W., Rezak, R., and Ginsburg, R. N., (1964). Classification and environmental significance of algal stromatolites. *J. Geol.* **72,** 68–83.

Logan, B. W., Harding, J. L., Ahr, W. M., Williams, J. D., and Snead, R. G. (1969). Carbonate sediments and reefs, Yucatán Shelf, *Mem.—Am. Assoc. Pet. Geol.* **11**.

Longman, M. W. (1981). A process approach to recognise facies of reef complexes. *Spec. Publ.—Soc. Econ. Paleontol. Mineral.* **30,** 9–40.

Longman, M. W. (1982). "Carbonate Diagenesis as a Control on Stratigraphic Traps (with examples from the Willington Basis)," Course Note Ser., No. 21. Am. Assoc. Pet. Geol., Tulsa, Oklahoma.

Macintyre, J. G. (1977). Distribution of submarine cements in a modern Caribbean fringing reef, Galeta point, Panama. *J. Sediment. Petrol.* **47**(2), 503–516.

McKerrow, W. S., ed. (1978). "The Ecology of Fossils; An Illustrated Guide." MIT Press Cambridge, Massachusetts.

Majevske, O. P. (1969). "Recognition of Invertebrate Fossil Fragments in Rocks and Thin Sections," Int. Sediment. Petrogr. Ser., No. 13. Leiden, Brill.

Maxwell, W. G. H. (1968). "Atlas of the Great Barrier Reef." Elsevier, Amsterdam.

Mazzullo, S. J. (1980). "Stratigraphic Traps in Carbonate Rocks," Repr. Ser., No. 23. Am. Assoc. Pet. Geol., Tulsa, Oklahoma.

Milliman, J. D. (1974). "Marine Carbonates. Recent Sedimentary Carbonates. 1." Springer-Verlag, Berlin and New York.

Mimram, Y. (1977). Chalk deformation and large-scale migration of calcium carbonate. *Sedimentology* **24,** 333–360.

Monty, C., ed. (1981). "Phanerozoic Stromatolites. Case Studies." Springer-Verlag, Berlin and New York.

Murray, R. C. (1960). Origin of porosity in carbonate rocks. *J. Sediment. Petrol.* **30,** 59–84.

Neumann, A. C., and Land, L. S. (1975). Lime-mud deposition and calcareous algae in the bight of Abaco, Bahamas: A budget. *J. Sediment. Petrol.* **45,** 763–786.

Orme, G. R. (1977). Aspects of sedimentation in the coral reef environment. *In* "Biology and Geology of Coral Reefs" (O. A. Jones and R. Endean, eds.), IV Geology II, pp. 129–182.

Orme, G. R., and Brown, W. W. (1963). Diagenetic fabrics in the Avonian limestone of Derbyshire and North Wales. *Proc. Yorkshire Geol. Soc.* **36,** 51–66.

Park, W. C., and Schott, E. H. (1968). Stylolites, their nature and origin. *J. Sediment. Petrol.* **38,** 175–191.

Peryt, T. (1983). "Coated Grains." Springer-Verlag, Berlin and New York.

Pettijohn, F. J. (1975). "Sedimentary Rocks," 3rd ed. Harper & Row, New York.

Plumley, W. J., Risley, G. A., Graves, R. W., Jr., and Kaley, M. E. (1962). Energy index for limestone interpretation and classification. *In* "Classification of Carbonate Rocks" (W. E. Ham, ed.), Mem. No. 1, pp. 85–107. Am. Assoc. Pet. Geol., Tulsa, Oklahoma.

Pratt, B. R. (1984). "*Epiphyton* and *Renaleis*—Diagenetic microfossils from calcification of coccoid blue-green algae. *J. Sediment. Petrol.* **54,** 948–971.

Pray, L. C., and Murray, R. C., eds. (1965). "Dolomitization and Limestone Diagenesis—A Symposium," Spec. Publ., No. 13. Soc. Econ. Paleontol. Mineral. Am. Assoc. Pet. Geol., Tulsa, Oklahoma.

Purdy, E. G. (1968). Carbonate diagenesis: An environmental survey. *Geol. Rom.* **7,** 183–228.

Purdy, E. G. (1974). Karst-determined facies patterns in British Honduras: Holocene carbonate sedimentation model. *Am. Assoc. Pet. Geol. Bull.* **58,** 825–855.

Purser, B. H. (1973). "The Persian Gulf. Holocene Carbonate Sedimentation and Diagenesis in a Shallow Epicontinental Sea." Springer-Verlag, Berlin and New York.

Railsback, L. B. (1984). Carbonate diagenetic facies in the upper Pennsylvania Dennis Formation in Iowa, Missouri, and Kansas. *J. Sediment. Petrol.* **54,** 986–999.

Read, J. F. (1985). Carbonate platform facies models. Am. Assoc. Petrol. Geol. Bull. 69, pp. 1–21.

Read, J. F., and Grover, G. A., Jr. (1977). Scalloped and planar erosion surfaces, middle Ordovician limestones, Virginia: Analogues of Holocene exposed karst or tidal rock platforms. *J. Sediment. Petrol.* **47,** 956–972.

Reading, H. E., ed. (1978). "Sedimentary Environment and Facies." Blackwell, Oxford.

Reeckmann, A., and Friedman, G. M. (1982). "Exploration for Carbonate Petroleum Reservoirs" (Elf-Aquitaine Centre de Recherches de Boussens et de Pau). Wiley, New York.

Reijers, T. J. A. (1972). Facies and diagenesis of the Devonian Portilla Limestone Formation between the river Esla and the embalse de la Luna, Cantabrian Mountains, Spain. *Leidse Geol. Meded.* **47,** 163–249.

Reijers, T. J. A. (1980). Sedimentary mechanism in Spanish Devonian carbonates. *Geol. Mijnbouw* **59,** 87–96.

Reijers, T. J. A., and Bartok, P. (1985). Porosity characteristics in fractured Cretaceous carbonate reservoirs, Vla Pas field area Maracaibo basin, Venezuela. *In* "Carbonate Petroleum Reservoirs," (P. O. Roehl and Th. W. Chogneth, eds.), pp. 407–424. Springer-Verlag, Berlin and New York.

Reijers, T. J. A., and Ten Have. A. H. M. (1983). Ooid zonation as indication for environmental conditions in a Givetian-Frasnian carbonate shelf-slope transition. *In* "Coated Grains" (P. Peryt, ed.), pp. 197–207. Springer-Verlag, Berlin and New York.

Reineck, H. E., and Singh, I. B. (1973). "Depositional Sedimentary Environments." Springer-Verlag, Berlin and New York.

Rich, J. L. (1951). Three critical environments of deposition, and criteria for recognition of rocks deposited in each of them. *Geol. Soc. Am. Bull.* **62.**

Roehl, P. O., and Choquette, P. W., eds. (1985). "Carbonate Petroleum Reservoirs." Springer-Verlag, Berlin and New York.

Rubin, D. M., and Friedman, G. M. (1977). Intermittently

emergent shelf carbonates: an example from the Cambro-Ordovician of eastern New York State. *Sediment. Geol.* pp. 81–106.

Schlager, W., and Bolz, H. (1977). Clastic accumulation of sulphate evaporites in deep water. *J. Sediment. Petrol.* **47**(2), 600–610.

Scholle, P. A. (1978). "A Color-Illustrated Guide to Carbonate Rock Constituents, Textures, Cements and Porosities," Mem. No. 27. Am. Assoc. Pet. Geol., Tulsa, Oklahoma.

Scholle, P. A., Arthur, M. A., and Ekdale, A. E. (1985). Pelagic Environments. *In* "Carbonate Depositional Environments," (P. A. Scholle, D. G. Bebout, and C. H. More, eds.) Am. Assoc. Petrol. Geol. Mem. 33 pp. 539–618.

Seilacher, A. (1964). Biogenic sedimentary structures. *In* "Approaches to Paleoecology" (J. Imbrie and N. D. Newell, eds.), pp. 269–315. Wiley, New York.

Shaver, R. H. (1977). Silurian reef geometry—new dimensions to explore. *J. Sediment. Petrol.* **47,** 1409–1424.

Shinn, E. A. (1969). Submarine lithification of Holocene carbonate sediments in the Persian Gulf. *Sedimentology* **12,** 109–144.

Stearn, C. W., Scoffin, T. P., and Martindale, W. (1977). Calcium carbonate budget of a fringing reef on the west coast of Barbados. Part I. Production and productivity. *Bull. Mar. Sci.* **27**(3).

Thornton, S. E., Pelkey, O. H., and Lynts, G. W. (1978). A lagoonal crustose coralline algal micro-ridge: Bahiret el Bibane, Tunesia. *J. Sediment. Petrol.* **48,** 743–750.

Toomey, D. F., ed. (1981). "European Fossil Reef Models," Spec. Publ., No. 30. Soc. Econ. Paleontol. Mineral., Tulsa, Oklahoma.

Turmel, R., and Swanson, R. (1964). "South Florida Carbonate Sediments" (R. N. Ginsburg, ed.), Sedimenta 2, pp. 26–33. Comp. Sediment. Lab., University of Miami, Coral Gables.

von Gümbel, C. W. (1888). "Geologie von Bayeren. I. Grundzüge der Geologie." Fischer, Kassel.

Walkden, G. M. and Berry, J. R. (1984). Syntaxial overgrowths in muddy crinoidal limestones: cathodoluminescence sheds new light on an old problem. *Sedimentology* **31,** 251–267.

Walker, R. G., ed. (1979). "Facies Models," Geosci. Can. Reprint Ser., No. 1. Geol. Assoc. Can., Toronto.

Walter, M. R., ed. (1976). "Stromatolites," Dev. Sedimentol., No. 20. Elsevier, Amsterdam.

Walther, J. (1894). "Einleitung in die Geologie als historische Wissenschaft." Fischer, Jena.

Watts, N. R. (1977). Pseudo-anticlines and other structures in some calcretes of Botswana and South Africa. *Earth Surf. Processes* **2,** 63–74.

West, I. M. (1965). Macrocell structure and enterolithic veins in British Purbeck gypsum and anhydrite. *Proc. Yorkshire Geol. Soc.* **35,** 47–57.

Wilson, J. L. (1975). "Carbonate Facies in Geologic History." Springer-Verlag, Berlin and New York.

Winkler, H. G. F. (1967). "Petrogenesis of Metamorphic Rocks," 2nd ed. Springer-Verlag, Berlin and New York.

Winland, H. D. (1969). Stability of calcium carbonate polymorphs in warm shallow seawater. *J. Sediment. Petrol.* **39,** 1579–1587.

Zenger, D. H., Dunham, J. B., and Ethington, R. L., eds. (1980). "Concepts and Models of Dolomitisation," Spec. Publ., No. 28. Soc. Econ. Paleontol. Mineral., Tulsa, Oklahoma.

Alphabetical Indices

ENGLISH

DUTCH

A

B

C

L

M

N

O

T

U

V

GERMAN

K

SPANISH

FRENCH

D

E

F

COMPUTER COMPATIBLE CODES

COMMONLY USED (INFORMAL) ABBREVIATIONS